天开
书系

a vertical forest /

instructions booklet for the prototype of a forest city

stefano boeri 斯坦法诺·博埃里 编著

一座垂直的森林

森林城市原型的说明书

edited by 编辑
Guido Musante 圭多·穆桑特
Azzurra Muzzonigro 阿族拉·穆重尼格罗

with the contribution of 协助
Michele Brunello 米歇尔·布鲁奈洛
Laura Gatti 劳拉·加蒂
Julia Gocałek 尤丽亚·高卡特赖克

Translated by 翻译
Yibo Xu 胥一波

同济大学出版社
TONGJI UNIVERSITY PRESS

图书在版编目（CIP）数据

一座垂直的森林 /（意）博埃里（Boeri,S.）编著；胥一波译 . —上海：同济大学出版社，2016.4

书名原文：BOSCO VERTICALE

ISBN 978-7-5608-6284-2

Ⅰ. ①一… Ⅱ. ①博… ②胥… Ⅲ. ①高层建筑－建筑设计 Ⅳ. ① TU972

中国版本图书馆 CIP 数据核字（2016）第 071696 号

天开书系

一座垂直的森林

斯坦法诺 · 博埃里（Stefano Boeri）编著，胥一波 译

天开书系 顾问：陈友祥
天开书系 策划：陈彼岸

责任编辑：吕炜
责任校对：徐春莲
排版制作：完颖

出版发行　同济大学出版社 www.tongjipress.com.cn
（上海市四平路 1239 号　邮编：200092　电话：021-65985622）
经　销　全国各地新华书店、建筑书店、网络书店
印　刷　上海丽佳制版印刷有限公司
开　本　787mm×1 092 mm　1/16
印　张　10.25
字　数　205 千字
版　次　2016 年 4 月 第 1 版　　2016 年 4 月 第 1 次印刷
书　号　ISBN 978-7-5608-6284-2
定　价　98.00 元

Foreword 序

I have read many works of architectural history. Some are written by Chinese authors, some by foreigners. Some are talking about classical architecture, some about modernism. I developed a conception of how cities and buildings are fostered in different societies, eras, economies and culture in the journey through books, which is usually later materialized in journeys through the world. What a fabulous experience it is! Buildings are an indispensable element of culture. Where there are humans, there are houses to shelter them from winds and rains, and markets for them to exchange goods. Naturally, these markets, if large enough, will eventually develop into cities. As an old saying goes, it takes only one year to make a village, two a city and three a metropolis, the later of which is how Chengdu, the capital of Sichuan province, got its name (in Chinese, 'cheng' means emerge and make, and 'du' means metropolis). In all forms of societies, business is the most effective impetus of prosperity. While the number of markets climbs and their scales expand, there will be more kinds of commodities available and the region will get more affluent. The resulting growth of population and city will translate into more houses, which will gradually get their own lives and evolve with the time passing. Then, we demand more materials from the earth to explore the various possibilities of buildings: as we initially began digging down, hides, grass and wood, stone and soil were our primitive 'blocks' of building. Later, we dug deeper and turned to bricks, steel, cement, glass and all kinds of metal, which were synthesised, manufactured and applied to buildings. So, besides the cause of personal security, buildings have gradually gained the function of prompting prosperity, carrying belief, and engraving down civilisation, furthering our exploitation and perception of the world. As wars and businesses help to blend civilisations of different regions, the human society has always been advancing. While wandering in different cities and countries, one can feel clearly that all men are vulnerable to the inevitable end, but those marvelous buildings will forever stay, recording the past and awaiting the future.

我阅读过很多个版本的建筑史著作，中国人写的、外国人写的、古典的、近现代的，想象不同的社会背景、时代环境、经济状况和文化风格造就了各种不一样的城市和建筑群落，在后来的旅行中又能一一见证，实在是一种很棒的体验。

建筑是个绕不过去的文化元素，有人的地方就一定有遮风避雨的房子，人一旦汇集，很快就会出现交换商品的集市，规模大了自然成了城市，有句话叫：一年成聚，二年成邑，三年成都。家乡首府蓉城的名字也来源于此。在所有的社会形态里，商业都是最快的繁荣催化剂，交易的市集增多变大，商品种类愈加丰富，一个地区会逐渐富裕，人口上升，城市的规模逐渐扩大，房子就多了起来，随着历史推进还慢慢开始有了生命，一点一点开始进化。我们向土地索取更多的材料来尝试更多的建筑可能性，从地表开始一直向地下挖掘，兽皮、草木、石土等原始素材成为建筑物的血肉，再继续掘土烧砖瓦、采矿铸钢铁，水泥、玻璃、金属等更多的材料被发现、合成、制造并使用到建筑上。建筑的功能从保护我们的安全开始，慢慢地孕育了繁荣，承载了信仰，镌刻了文明，帮我们不断地去开发和认知世界。战争和商业使不同地区的文明融合，人类社会不断推进和演化，在不同国家的城市里漫步，会深深地感受到这些，人将逝去，而那些伟大的建筑会留下，记录过去，也在期待未来。

Foreward

Records about economy, art, industry and other disciplines can well reflect the change of buildings through ages. In modern times, while both architecture and social reform leap forward, our buildings rise high into the sky and roads spread toward the sea; meanwhile, old buildings collapsed, but more new ones were springing up. It seems that buildings all have their own lives, growing from short, small and simple ones with rigid facades into tall, large and complex ones with curved profiles. Nonetheless, as more mega steel "towers" shoot up into the sky and more concrete roads spread under our feet, biological species are disappearing, the sky and the sea are losing their bright colour, and the environment is deteriorating and becoming more and more alien to all of us. Suddenly, it comes to us that human civilisation has diverted from its original course of development. Every single step forward by human means one step back by the nature. Now, it is a critical point for the development of cities, where the comfort has sharply declined. Of course, "man-built environment" can never match the work of nature. It is impossible for us to demolish cities and disperse humans so as to restore nature, but no communities should develop based on destruction and sacrifice. Fortunately, it seems that we have found some possibilities in buildings: some real brave hearts and great minds are thinking about our future, and their beautiful depictions of the future should be turned into reality. I believe, their inspirations will develop over time, helping to clean our sky, bringing back butterflies and birds, contributing to city's biodiversity and enriching our life.
The road for answers and solutions is bound to be bumpy, and the result unknown, but we would go hand in hand for a simple reason: here is our home.

Akpher
March 2016, in Milan

再读经济、艺术、工业等各领域的历史典籍，都可一一印证各个时代的建筑，步入近现代以后，建筑的发展与社会变革一样突飞猛进，我们往天空伸展，我们往海上筑路，老去的建筑崩塌，而更多的新建筑在崛起。建筑已经有了自己的生命，从矮到高，从小到大，从简单到复杂，从直线到曲线。但是当太多的钢铁巨塔插向云端，水泥路面铺满脚下的时候，一个个物种开始消失，天空和海洋开始褪色，环境越来越糟糕和陌生。我们突然发现文明的进程似乎已经偏离了航向，人类在迈步向前，而大自然退向地平线的远方。在当今，城市的发展模式抵达临界点，舒适度生存度急剧下降，人造的环境又怎么比得过上帝的作品。我们不可能拆毁城市驱散人群来返还大自然的原貌，但任何族群的进步发展，都不应该以破坏和牺牲为前提。

不过幸而我们似乎找到一些可能性，在我们中间总有一些真正的勇士和智者在思考未来，而这些对未来的美丽描绘就应该被一一呈现。我相信这些闪亮的灵感经过时间的孕育和发酵，一定可以让蓝天回归，让蝴蝶和鸟儿回归，让城市的生命重新丰富起来，让生活的颜色丰富起来。

寻找答案的前行之路崎岖坎坷，且未知成败，但我们愿意一起携手尝试，只因为，这里，是我们的家园。

天开书系策划

2016 年 3 月写于米兰

seven inspirations / 七个灵感

introduction

by / stefano boeri

斯坦法诺·博埃里

one

In 1972 I was 16 years old and I had no idea that travelling around the streets of my city,
HUNDERTWASSER, FRIEDENSREICH * P. 78
Milan, there was a bizarre character – the Austrian artist Friedensreich Hundertwasser – who while holding a tree was preaching the idea of a new style of architecture, built around the presence of trees in houses, courtyards and rooms.

In the middle of the streets with a small oak tree in his hand and just a few metres from the La Scala Opera House, Hundertwasser was making the case for organic architecture, based on a standard that governed the relationship between the number of humans
HUMANS * P. 78
and the number of trees in any space where people lived.

At that time I was a high school student involved in the movements of the extra-parliamentary left and I was walking the same streets protesting in Milan – not about issues such as ecology and environmental sustainability but rather the great problems of social inequality, the right to education, of "imperialism".

I considered that ecology and environment issues were superfluous and irrelevant, typical concerns of the "capitalist bourgeoisie".

Yet in Italy, in Florence during those same years, there were some young architects and artists in their twenties and thirties who were committed to the culture of protest. They formed the Gruppo 9999 and began to think about the relationship between trees
BASIC RADICALITY * P. 54
and humans in cities and build extreme and radical visions of the urban future, images of urban settings filled with forests and woods permeated by architecture.

But even their vision, so strong and radical and disturbing, fell foul of the indifference of the dominant culture within the Italian and European intellectual left.

I never thought that 40 years later I myself would be the author, right here in Milan,
(ANTI-) ANTICITY * P. 51
of an architecture that aims to revolutionize the relationship between trees and humans
CULTURAL CELL * P. 65
in an urban centre and which aims to promote a new idea of the city.

1972年，我16岁。那时候的我，并没有意识到，在米兰城，我的城市街道上，出现了一位怪异人物的踪迹。他就是奥地利艺术家：百水先生。他手握着一颗树苗，宣扬着一个前所未有的建筑理念：围绕着树木，建造房屋、庭院和房间。

百水先生 * p.78

在马路中央，离斯卡拉歌剧院仅几步之遥，百水先生手握着一株橡树苗。那时他正在进行一个关于有机建筑的实验。这个项目试图在所有的人居空间中建立一个平衡人口数量和树木数量关系的标准。

人类 * p.78

那时的我，还是一个高中生，也在米兰的同一条街道上，在国会门口，参与左派社会运动的示威活动——但那时，我并不是为了生态和环境的可持续性，确切地说，是为了社会不平等、教育权力和“帝国主义”那一类更重要的问题在抗争。

那时的我以为，生态和环境问题是多余的、没有时代意义的。它们属于典型的“资产阶级”才关心的问题。

然而，在那几年里，意大利的佛罗伦萨，有一些二三十岁的年轻建筑师和艺术家，把生态与环境作为人类文化，并投身其中。他们组建了9999团，开始思考城市中树与人之间的关系。同时，他们还对城市的未来作出了激进的展望，想象着森林和树木，渗透在城市建筑之间。

激进的原型 * p.54

然而，即使他们的愿景如此强烈和震撼人心，却与当时的意大利和欧洲主流文化格格不入。

我亦未曾料到，40年后，就在米兰城中心，我自己会创作出一个变革了城市中树木与人类关系的建筑作品，并致力投身于推广这场变革。

反－反城市 * p.51
文化细胞 * p.65

two

BIG TREE * P. 55

The idea of building a tower completely surrounded by trees came to me in early 2007 in Dubai – one of the cradles of the new oil and financial capitalism – when as editor-in-chief of "Domus" I was following the frantic construction of a city in the desert consisting of dozens of new towers and skyscrapers.

All clad in glass or ceramic or metal.

All reflecting the sunlight and therefore heat generators: in the air and especially on the ground, the area inhabited by pedestrians.

At that time I was teaching at the Graduate School of Design at Harvard and the School magazine ("Harvard Design Magazine" [1]) had published a piece of research by Alehandro Zaera Polo which explained that 94% of the tall buildings in the world built after 2000 were covered in glass.

MINERAL CITY * P. 87

Glass and mineral skins in an increasingly artificial and mineral city.

At that time I was starting the design of two towers in the centre of Milan and suddenly – the most radical and bizarre ideas come without warning – it occurred

MATERIALS * P. 85

to me to create two eco-friendly towers; two towers covered not in glass but in leaves – leaves of plants, shrubs, but especially the leaves of trees.

Two towers covered in life.

To convince my clients – the Italian branch of a multinational American real estate company – I asked a journalist friend to publish a picture in an Italian newspaper showing the two towers covered with trees and a compelling title: "the first ecological

DEMINERALIZATION * P. 65

and sustainable tower is going to be created in Milan".

[1] *High-Rise Phylum*, on: "Harvard Design Magazine" n. 26, Spring/Summer 2007.

二

2007 年初，在新兴石油与金融资本的摇篮——迪拜，我突然有了一个想法：建造一栋四周被树木环绕的塔楼。当时，作为 *Domus* 杂志的主编，我正在跟踪报道这座正疯狂建设摩天大楼的沙漠城市。所有的建筑表皮都由玻璃、陶瓷材料或金属构成。所有的建筑都反射着阳光，成为一个个巨大的加热器，炙烤着空气，炙烤着地面上的行人。

大树 * p.55

当时，我在哈佛大学设计研究院任教。扎埃拉 · 波罗在哈佛的校刊（哈佛设计杂志第一期 [1]）上发表了一份研究报告，其中提到，2000 年后新建的高楼中有 94% 的建筑外立面被玻璃覆盖。

矿物城市 * p.87

玻璃表皮、金属外观及人造矿物立面的建筑，在城市中越来越普遍。

那时，我正在设计米兰市中心的两栋塔楼。毫无征兆地，一个冲动怪异的想法产生了——创造两栋生态友善的塔楼；他们不被玻璃，而是被树叶覆盖，不仅是灌木的树叶，更重要的是乔木的树叶。

材料 * p.86

生机勃勃的两栋大楼！

去矿化 * p.66

为了让客户，一家美国跨国地产的意大利分公司支持这个创意，我邀请了一位记者朋友，在意大利的报纸上发表了一篇文章，题为《首栋生态建筑将在米兰建成》，并附上了一张覆盖着绿色植物的两栋大楼的图片。

[1] 高层之门．哈佛设计杂志．2007 年春／夏．26 期．

HEIGHTS * P. 76
In fact, the promise and the intention of these two buildings (120 and 90 metres high respectively) was to noticeably reduce energy consumption thanks to the filter that
MATERIALS * P. 85 MICROCLIMATES * P. 86
a facade of leaves exerts on the sunlight plus the microclimate that is created on the
BALCONIES * P. 52
balconies, which reduces the difference in temperature between the inside and the outside of the apartments by about 3 degrees.

I added in that article - which was so successful as to push my clients to take this little 'quirk' seriously - that in addition to carbon dioxide, the leaves of the trees would also absorb the pollutant micro-particles created as a result of urban traffic and so would
SUSTAINABILITY * P. 97
help clean the air in Milan, as well as producing oxygen in turn.

In the following months, together with the architects in my studio [2] we wrote a 'Manifesto for the Vertical Forest' which promoted the idea of a living and sustainable architecture that would reduce fuel consumption and therefore the human impact on the environment.

The truly revolutionary aspect of the project was not of course the presence of trees
URBAN FOREST * P. 101
and shrubs on the balconies; but the idea of hosting nearly 800 trees from 3 to 9
POTS * P. 91
meters tall along the kilometer and 7000 metres of pots that lined the perimeter of the balconies of the two towers.

The idea was to have two trees for every inhabitant of the two towers, leading to a total of 21,000 plants (4,000 shrubs and 15,000 perennials and climbers). Effectively, it was the idea of building a tower for trees - which incidentally housed human beings.

[2] Boeri Studio, founded in 1999 and active until 2008 was formed by the three partners Stefano Boeri, Gianandrea Barreca and Giovanni La Varra.

高度 * p.76 材料 * p.86 阳台 * p.53 微气候 * p.86

实际上，建设这两栋大楼（分别为 120m 和 90m 高）的初衷，是希望通过树叶在光照条件下的过滤作用，优化阳台的小气候，从而显著降低能量消耗。这样可以将室内外温差减少大约 3℃。

可持续性 * p.97

那篇报道卓有成效，以至于我的客户幡然悔悟，开始认真而快速地思考了这个创意的潜在价值——除了降低二氧化碳排放量以外，树叶还可以吸收由城市交通产生的污染物微粒子，进而净化米兰城的空气并产生氧气。

在接下来的几个月里，工作室 2 的建筑师们共同编写了《垂直森林宣言》，倡议宜居的生态建筑，降低燃料消耗量，减少人类活动对环境造成的影响。

城市森林 * p.101 种植盆 * p.91

这个项目真正变革性的特点，当然不仅仅是把乔木和灌木呈现在阳台上，而是在这两栋大楼上种植了大约 800 株 3～9m 高的植物。为种植这些植物，阳台上的种植盆总周长达到了 7km。

这个想法的初衷是为每位住户提供 2 株植物，最终达到引入 21 000 株植物（其中包括 4 000 株灌木和 15 000 株多年生植物和攀援植物）的目标。实际上，可以认为，这是两栋为树木而建的大楼，只不过凑巧人类也在上面居住。

2 博埃里工作室，1999 年成立，三个合伙人为博埃里、巴莱卡和拉瓦拉。

three

The trees are not “green”, they are not “forest”, they are not “nature”.
PLANTS * P. 90
Every tree is a character in the life-giving story of the planet, with its own biography and a mysterious ability to preserve our public and private memories. I owe my
THE BARON IN THE TREES * P. 98
obsession with trees to Cosimo Piovasco di Rondò, the little Baron who one evening in 1767 in Ombrosa, a small town in western Liguria, decided – at the age of 12 – to leave the ground and live in trees for the rest of his life. The character from the novel by Italo Calvino, published in 1957, is a staple in the imagination of my adolescence; to him I owe my fascination with the forests of olive trees and oaks that line the shores of the Mediterranean and their undergrowth of juniper, myrtle and helichrysum. Also to the Baron of Ombrosa, born just a few kilometers from Badalucco, the village which is home to the roots of my father’s family, I also perhaps owe a taste for obstinacy in radical and irreversible choices. But there were other memories from my life that inspired my obsession with trees, such as visits to the building site of a small
THE HOUSE IN THE WOODS * P. 99
house designed in 1968 by my mother Cini Boeri in the woods of Osmate, near Lake Maggiore. I was 12, like Cosimo, and I remember the decision (then very much against the mainstream) of building a house that took up land between the birch trees without cutting down any of them. A zigzag house, with inset sections built around the trees and big windows that looked out on the branches. A few years later, in 1972 – the very year
HUNDERTWASSER, FRIENDENSREICH * P. 78
in which Friedensreich Hundertwasser was walking in the streets of Milan with a tree
CELENTANO, ADRIANO * P. 60
– a great Italian singer and artist, Adriano Celentano, wrote one of his most beautiful songs about the risks of pollution and land speculation – “A 30 stories-high tree” – which ends with the vision of a tree that grows up for thirty stories in the middle of the city [3]. With intuitive and powerful simplicity, Celentano had opened the imagination to a new architecture.

[3] “... you shouldn’t grouse if the concrete blocks your nose, neurosis is in fashion: if you haven’t got it you’re out. Ouch. I cannot breathe any more, I feel that I’m choking a bit, I feel my breath going down, it goes down and doesn’t come back up, I only see that something is emerging... Maybe it’s a tree yes it’s a tree 30 floors high”.

三

树木并不是“绿色”，也不是“森林”，亦不是“自然”。

植物配置 * p.90

每一棵树，都在星球的生命故事中扮演着一个角色，它们有着自己的传奇，还具备保存公众和个体记忆的神秘能力。我对树木的痴迷源于柯西谟，他是一位来自西利古里亚地区欧布鲁萨小镇的男爵。1767 年，在他 12 岁的一个夜晚，男爵毅然离开地面，决定在树上度过余生。柯西谟是意大利作家卡尔维诺 1957 年发表的小说中的角色，他也是我青春期想象的主角。我着迷于地中海海滨沿岸的橄榄树和橡树森林，着迷于它们树下的杜松树，着迷于桃金娘和蜡菊，而这些，都源于他。男爵出生的欧布鲁萨，距离我父亲家族的根基巴达卢科村庄只有几公里。这也许是我既固执己见，又特立独行的原因之一。当然，生命中还有其他的一些记忆，也是我痴迷于树木的缘由。比如，参观我母亲奇尼 · 博埃里于 1968 年在马焦雷湖附近的奥斯马泰森林设计的小房子。我那时 12 岁，和柯西谟当年一般大。我记得，她决定在桦树之间建造房子，但绝不砍伐树木，当时这种做法十分违背主流 。于是，一栋 Z 字形的建筑诞生了，它行走穿梭在林间，一旦打开大窗户，窗外枝叶扶疏。几年后，也就是 1972 年——百水先生拿着树苗走在米兰大街上的那年，伟大的意大利歌手和艺术家阿德里亚诺 · 切伦塔诺，写下了一首他最美的歌《30 层高的树屋》，抗议污染与土地投机。在歌曲的结尾描述了一幅美景——城市中心有一棵长到 30 层楼高的树 [3]。凭借本能和直白的强大力量，切伦塔诺开启了新式建筑的灵感之窗。

《树上的男爵》* p.98
森林之屋 * p.99
百水先生 * p.78
阿德里亚诺·切伦塔诺 * p.60

[3] “如果混凝土阻挡了你的嗅觉，不要有怨言，神经官能症正在流行。如果你还没有受到感染，那你太落伍了。喔，我简直无法呼吸了，我感觉有点窒息，呼吸越来越困难，越来越困难，无法恢复，我只看到某种东西在显现…… 或许是一棵树，哦是的，那是一棵 30 层楼高的树。”

four

The prototype of a Vertical Forest – the two Milan towers – is today inhabited by 380 HUMANS * P. 78 human beings and 780 trees, in addition to nearly 5,000 shrubs and several thousand climbing plants and perennials. And of course, by different species of birds that nest at all heights.

It took months of research and experiments conducted with a group of outstanding experts in botany, ethology and SUSTAINABILITY * P. 97 sustainability [4], to solve problems that architecture had never before had to deal with: how to prevent a tree being broken by the WIND TUNNEL * P. 103 wind and falling from a height of 100 meters; how to ensure MAINTENANCE SYSTEM * P. 84 continuous and precise IRRIGATION SYSTEM * P. 80 watering of trees planted at heights where conditions of humidity and exposure to sun are very different; how to prevent the life of the trees being jeopardized by the personal choices of the owners of the apartments.

I also owe a great deal to the courage of my clients [5] not just for the significant investment in this quest and exploration for new ideas and solutions, but also the sharing of the risks involved in creating the prototype of a new MUTE ARCHITECTURE * P. 87 dimension of architecture.

[4] We would never have been able to design and build the Milan Vertical Forest without the contribution of Laura Gatti, who with Emanuela Borio designed the green project and day by day followed the selection of species, the study of the best living conditions for plants and their maintenance, without the technical and structural solutions designed by Arup Italy to support the weight of the earth on the perimeter of the balcony borders and for fixing the tree roots to the base of the pots and without the availability of the Peverelli company, particularely of the unforgettable Giorgio Peverelli, in terms of the procurement and quality of species of trees, shrubs and plants housed in the Vertical Forest.

[5] In particular to Manfredi Catella, CEO and Co-Founder of COIMA Sgr and his Team of technicians and managers.

人类 * p.78

垂直森林的原型——两栋米兰的高塔，如今已有 380 人和 780 棵乔木入住，另外还包括近 4 500 棵灌木和几千株攀爬植物与多年生植物。当然，在这里不同的高度上，还有不同鸟类的栖居之所。

可持续性 * p.97

一群出色的植物学、动物行为学和可持续发展方面的专家，花了几个月的时间来进行研究和实验 [4]，解决了建筑业之前无需考虑的问题。比如，如何预防树木被风摧毁，或从 100m 的高空坠落；如何确保，在不同高度、不同湿度和不同光照条件下，树木能获得持续而精确的浇灌；如何保护树木的生命，不被房屋所有者的个人选择所危害。

风洞 * p.103
养护系统 * p.85　灌溉系统 * p.80

很大程度上，我的勇气源于我的客户 [5]。不仅仅是因为他们对这个创意项目和方案的慷慨投资，他们还共担了开发一个新式建筑原型所存在的风险。

沉默建筑 * p.87

[4] 如果没有劳拉 · 加蒂的帮助，或许我们根本就不可能设计并建设米兰垂直森林。劳拉 · 加蒂与埃曼奴埃拉 · 博里奥一起设计了项目的绿化工程，每天都在跟进物种的配置，研究最适宜的植物生长环境与养护方式。此外，奥雅纳意大利公司提供的技术和结构方案，使在阳台边界承受土壤与植物负荷，并稳固根系的要求得以实现。而垂直森林内所有植物的采购和品质保障，则归功于佩维莱立公司。

[5] 特别是柯意马的总裁和联合创始人曼福莱狄 · 卡戴拉，以及他的技术和管理人员团队。

five

BIODIVERSITY * P. 55

Thinking in terms of biodiversity means taking note of the fact that today Europe is one big city; and that in today's world, megacities and metropolitan areas have multiplied, so that the area they cover extends like a continuous blanket over coasts, valleys, and plains, devouring thousands of hectares of nature and agriculture. Immense urban surfaces in many parts of the world have now surrounded both the few large remaining natural areas and turned them into "theme parks", as well as the agricultural areas converted into large vegetable plots, or enclosed gardens, *hortus conclusus*.

The limitless growth of these endless cities – which means that by 2050 the urban population will exceed 70% of the global population – has altered the biological balances and has greatly reduced the biodiversity of species, on the one hand accelerating the reduction and extinction of some species of plants and animals,

THIRD LANDSCAPE * P. 100

and on the other hand forcing them to invade unfamiliar territories due to their forced expropriation from uncontrolled nature.

Deers that are forced to move into the centres of many mountain towns, foxes that now inhabit the suburbs and the underground of London, the wild boars that roam the streets of many cities in the Mediterranean are the indications of deep disturbances within the sphere of life, disturbances that signal the failure of a historic pact between nature and artifice; a disturbance that we cannot think of dealing with through returning to the past and re-presenting our city within the boundaries of the past.

BIODIVERSITY * P. 55

The great challenge of biodiversity must therefore be faced and won within the current

CONTINUOUS CITY * P. 64

conditions of the Continuous City, the infinite metropolis that surrounds us and in which we live.

五

生态多样性 * p.56

按照生物多样性的思路去思考，就意味着人们已经注意到如今的欧洲事实上已犹如一个大城市。当今，世界的大都市圈正成倍地增加面积，它们所覆盖的区域，绵延地覆盖在海滨、山谷和平原之上，吞噬着上万公顷的大自然和农耕用地。世界上很多地方的城市，把大面积的自然区域都变成了“主题公园”，而纯粹的自然空间则少之又少。与此同时，很多农业区转变成了较大的菜田、封闭式花园或种植庭院。

这些城市肆无忌惮地扩张，这意味着，到 2050 年城市人口将占全球总人口的 70%。由此，生物平衡被打破，生物多样性大大削弱。一方面，这会加速某些植物和动物减少和灭绝的速度；另一方面，由于被迫离开家园，这些生物必须被迫迁徙到陌生的区域生长。

第三景观 * p.100

鹿群被迫转移到许多山城的中心，狐狸则栖息于伦敦的郊区和市区，而野猪们在地中海的许多城市街道上转悠，这一切都是生态圈被严重扰乱的表现。这种干扰，标志着人类与自然界之间传承已久的平衡正在丧失。这种干扰，亦使得我们无力思考。该如何恢复旧貌？如何能不扩张，就能重新改造城市？

生态多样性 * p.56 连绵城市 * p.64

我们必须直面保护生物多样性这一巨大挑战。人们必须在连绵的城市群中，在我们所生活的大都市的现状下，赢得这场挑战。

It must be won with political tools: creating protected areas entrusted to the deep-rooted autonomy of nature, from which not only the presence, but also embedded examples of our species' governments and controls must be totally expunged.
In addition, agriculture should be planned in the territories around the metropolises and in large landlocked areas of European cities, a variety of crops and products need to be rediscovered and acquired, and not just a "grey" extension of fields planted with monocultures of cereal crops and inhabited by otters and crows.
But the great challenge of biodiversity can and must be fought and won thanks

DENSIFICATION * P. 66

also to architecture, through multiplying the places for the generation of plant biodiversity and wildlife within the denser and more congested urban areas, as well

DEMINERALIZATION * P. 65

as de-mineralizing urban surfaces, with green roofs, vertical gardens and real grafts of biodiversity, such as the Vertical Forest.
But this challenge, where it will not be possible to limit the pressures of urbanization,

(ANTI-) ANTICITY * P. 51

will be won only with the realization of genuine floral and faunistic cities, "Urban Forestry" where architecture is not just a frame or a focal point for nature, but which is created together with it, becoming inseparable.

要获胜，就必须通过行政手段去创造具有高度自治的自然保护区，不仅是去消除控制的存在，更是要彻底摆脱对于物种发展的控制与干涉。此外，在大都市周围的区域需要规划农业用地。在欧洲城市周围大面积的土地上，要去重新发现和获取多样化的作物和产品，决不能继续只种植单一粮食作物，这样的农田只有水獭和乌鸦才会于此栖居。

当然，我们也必须感谢建筑技术的发展，它使得生物多样性得以实现，正如垂直森林所做的，在稠密拥挤的市区增设植物多样性的繁殖地点，通过使用屋顶绿化、垂直花园和移植多样性生物对城市表面去矿化。

致密化 * p.66

批判性观点 * p.66

反一反城市 * p.51

虽然不可能减小城市化的压力，但是通过建设草木葱笼的城市，我们可以赢得生物多样性的成功。在“城市森林”中，建筑不再是存于自然界的一个构架或聚焦点，它应与大自然融为一体，不可分割。

HOWARD, EBENEZER * P. 77

In 1898 Ebenezer Howard published *A Peaceful Path to Real Reform*, a milestone text for a new pact between city and nature that was republished in 1902 in a new version with the title *Garden Cities of To-Morrow.*

Howard's proposal for addressing social inequalities, problems of pollution and traffic and risks of hygiene and public disorder resulting from the overwhelming urban growth linked to the processes of industrialization, was to design urban communities of 32,000 inhabitants around London and the larger cities. Cities with low density, with services and facilities for the community at the heart and around them a system of circles of homes surrounded by nature, able to combine the advantages of urban life and those of the countryside.

The first "Garden Cities" as examples and public property were built in 1903 in Letchworth and in 1920 in Welwyn and were the inspiring model for an important trend in planning and architectural thinking that ran through the 20th century, with names ranging from Lewis Mumford to Clarence Stein, Henry Wright to Clarence Perry and from Rexford Tugwell to Arthur Morgan.

URBAN FOREST * P. 101

A century later, the proposal for the creation of a worldwide system of "Urban forests" consisting of buildings that are homes to nature within their own structures is facing a different scenario, that of parts of the world where the urbanization of large numbers of peasants will for many years yet be an unstoppable process. This scenario sees agriculture – agriculture that is versatile and full of variety, finally able to produce food for the different urban social groups – again becoming a key resource for large metropolitan areas.

霍华德，埃比尼泽＊ p.77

1898 年，埃比尼泽 · 霍华德（1850—1928）发表了《通往真正改革的和平之路》。1902 年，他又更新出版了这一关于城市与大自然之间的契约的具有里程碑意义的文本，题名为《明日的田园城市》。

就工业化进程引起的城市增长过快而带来的社会不公平、污染与交通问题、卫生风险和公共混乱等现象，霍华德提议，在伦敦等大城市周围，设计可容纳 32 000 居民的城市社区。社区结合城市与乡村的优势，维持低密度，中心设置社区服务设施，房屋围绕社区服务中心而建，外围为自然环绕。第一批“田园城市”和公共地产分别建设于 1903 年的莱奇沃思和 1920 年的韦林。田园城市成为了风靡 20 世纪城市规划和建筑设计思潮的典范，受其影响的重要人物有刘易斯 · 芒福德（1895—1990），克拉伦斯 · 斯坦（1882—1975），亨利 · 莱特（1852—1908），克拉伦斯 · 佩里，雷克斯福德 · 特格韦尔（1891—1979）和亚瑟 · 摩根（1878—1975）。

一个世纪以后，想要在全球系统下再去创建一座由无数个容纳着自然之家的建筑构成的“城市森林”，又面临着全新的境况：在世界很多地方，城市化已成为不可逆的进程。这一背景下，因为多样化、多用途的农产品最终能够为城市的不同社会群体提供食物和原料，农业再次成为大都市圈的关键资源。

城市森林＊ p.101

This is a scenario which requires a strict reduction in the consumption of natural and agricultural soil produced by the continuous horizontal extension of urban areas with low building density, and that makes it increasingly difficult for local governments to deal with the social, economic and environmental costs of large city management.

ANTI-SPRAWL DEVICE * P. 52

In this scenario, the design of small high density "vertical cities" featuring an intensity of life that reduces the cost of managing energy and transport services by proposing a new balance between the urban, agricultural and natural spheres can become a significant opportunity.

But to combine the density given by a vertical growth of buildings and the biodiversity resulting from a new balance between nature and city, urban planning is not enough. The visions that generated large scale transformation processes of territories and areas were able to combine large scale city planning and perspective with the creation of individual and timely architectural devices, repeated throughout the territory.

REPLICABILITY * P. 92

So in the same way that the single family house with garden a century ago was the elementary module as expressed in Ebenezer Howard's Garden Cities scenario, the tree-tower or "vertical forest" could become in the coming years the device – repeated with endless variations – that will allow not only to engage ecosystems of biodiversity in the built city environment, but also to create a new form of city: the Urban Forest, a City/Forest where architecture neither binds nor restricts nature, but instead accepts it as an original component part.

BIODIVERSITY * P. 55

该设想要求，严格减少低密度的城市对自然土壤和农产品的消耗，阻止农业用地被城市化所占用。这使得当地政府，在平衡社会、经济和环境成本时困难重重。

反城市蔓延的设备 * p.52

在这种情况下，设计小型高密度的“垂直城市”，旨在倡导城市、农业和自然界之间的新平衡，形成一个可以降低能源管理和运输服务成本的有效集合。这是一个重要的机遇。

但是，想要将增长的建筑密度与自然界和城市平衡后所带来的生物多样性结合，仅仅依靠城市规划还是远远不够的。

可复制性 * p.92

要实现大规模区域变化的愿景，需要通过城市规划，还需要适时地与个体建筑创造实践相结合。

在一个世纪以前，带花园的独户住房就是霍华德田园城市的一个基本模块。而在未来的几年，“树之楼”或“垂直森林”，将成为变化无穷的应用策略——它不仅可以在城市环境当中成为生物多样性系统的一部分，而且还会创建一种新的城市形式：城市森林里的建筑，即不约束自然，也不限定自然，而是将自然环境作为其原始的组成部分。

生态多样性 * p.56

seven

In 1982 in front of the Fridericianum Museum in Kassel the German artist Joseph
BEUYS, JOSEPH * P. 54
Beuys built a triangle of 7,000 basalt stones, each of which was intended to be used for the planting of a tree.

Anyone paying a sum of money, could "adopt" one of the seven thousand stones; the money raised would be used to plant an oak tree.

So, day after day, the pile of stones diminished until it finally disappeared, and seven thousand new oak trees, each with one of the basalt stones at its base, appeared along the streets and avenues and in the squares of the city of Kassel.

HUNDERTWASSER, FRIEDENSREICH * P. 78
Like Friedensreich Hundertwasser, like the Florentine architects of the radical
BASIC RADICALITY * P. 54
movement, Joseph Beuys showed us the great challenge of the coming decades:
DEMINERALIZATION * P. 65
transforming rocks into trees means in fact transforming houses and streets into places inhabited by thousands of living species. It means imagining an architecture that does not "host" or "fence off" portions of nature, but which is created together with nature itself.

It means learning to live with trees, with their presence and their speed of growth, and with their extraordinary capacity, even in the most polluted and congested areas of the urban world, of accommodating and giving life to a wealth of species.

博伊斯，约瑟夫 * p.55

1982年，在卡塞尔的腓特烈博物馆前，德国艺术家约瑟夫·博伊斯用7 000块玄武石搭建了一个三角堆，每一块石头都是为了种一颗树。任何人只要付一笔钱，都可以“领取”7 000块石头中的一块，而每块石头所筹集的资金则用于种一棵橡树。随着时间一天天流逝，石头越来越少，最终消失。而在卡塞尔的街道和广场上，生长起了7 000棵小橡树——每一颗树的根，都握着一块玄武石。

百水先生 * p.78　激进的原型 * p.54

正如百水先生，正如激进运动中意大利佛罗伦萨的建筑师一样，约瑟夫·博伊斯向我们展示了未来几十年将面临的巨大挑战——将岩石转换为森林。这也就意味着将住房和街道转变成有上万生物物种居住的地方。你可以想象一下这样的建筑，它不“霸占”空间，不“隔离”自然，自诞生之日即融入自然。

去矿化 * p.66

这意味着，人们要学会与树木一起生活，见证树木的生长速度与容纳力，以及它们在调节自然方面无与伦比的能力。即使在那些城市里污染最为严重的地区和人口稠密的地区，也能为大量物种提供生存的环境。

tales from the vertical forest /

by / Stefano Boeri 文 / 斯坦法诺·博埃里

illustrations by / Zosia Dzierzawska

来自垂直森林的故事

插图 / 左西亚·泽鲁佳富苏卡

The writings "Tales from the Vertical Forest" are only fruit of the imagination and do not relate in any way to events, people and real places.

文章《来自垂直森林的故事》是虚构作品，不与任何真实事件、任何具体人物和地点存在关联。

The Parrots, the Smiths and the Strawberry tree

The first one arrived a year ago, with the orange breast and the black beak, then the others arrived. The Smiths' big Strawberry tree, on the 20th floor, is their living room. They meet there every day, especially in Autumn, and they talk, they talk; they have long conversations among the orange berries and the red flowers. The young Smiths, Luis and Linda, have given them names and they invent stories about them: Lucino, the one with the violet feathers, escaped from a bird shop. Lorella comes from the balcony on the third floor. Piero, the one with the yellow crest, the most talkative of all, from the parish church courtyard. Gina, beautiful and restless, from a faraway part of Milan. They were all in cages and now they are free. A flying family. With a thousand colours.

鹦鹉，史密斯和草莓树

一年前，一只有着橙色胸部、黑色喙的鹦鹉第一次造访。随后，它引来了它的伙伴们。

二十楼，史密斯家最大的草莓树，便是它们的起居室。

在那里，它们天天见面，尤其是在秋天。它们叽叽喳喳，互相交谈，在橙色莓果和红色的花朵之间长谈。

史密斯家年轻的路易斯和琳达，给这些鸟儿分别起了名字，并想象着他们的故事：

卢奇诺，拥有美丽的紫色羽毛，它刚从鸟店逃脱。

洛雷拉，来自三楼的阳台。

皮耶罗，有黄色的冠，是最健谈的，来自教区的院子。

吉娜，美丽绝伦又精力旺盛，来自米兰远郊。

它们以前都被关在笼子里，现在它们是自由的。

一个飞翔的家庭，五彩斑斓！

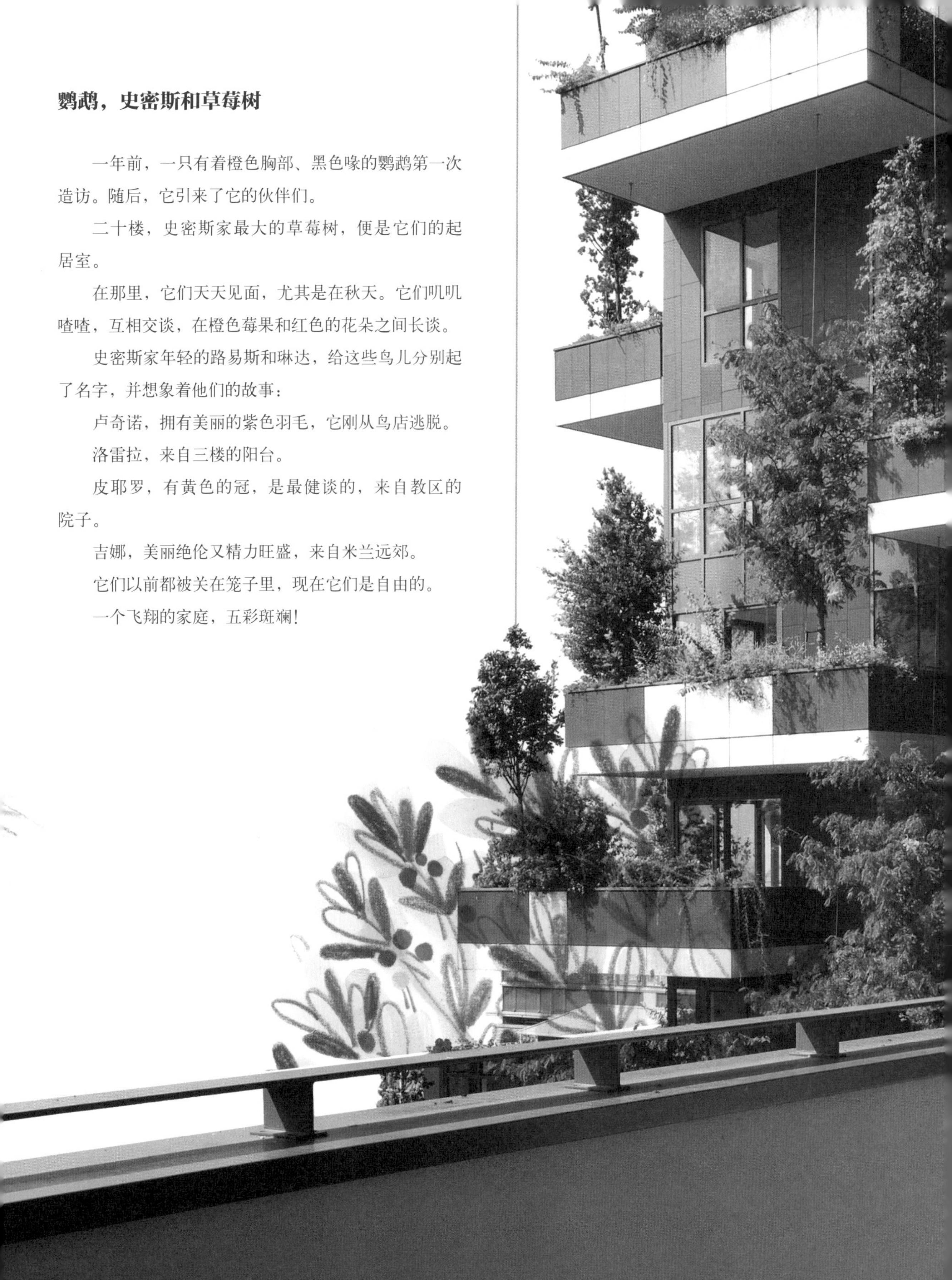

The Crows, the Bulgaronis and the Maple tree

Since we made the nest, everything has changed. Before, there was no way: they didn't want us. And just to keep us away, the Bulgaronis had tried everything; even those silly soft toys they tried to scare us with. But since our nest has been on their balcony, we have become an attraction. Together with Castaldis and Cerettis, who are on the floors above, they watch how our little ones grow. We see them as they look at us, hardly moving the curtains so as not to disturb us when we bring food. Now they have adopted us. We are the subject that has brought three families in the tower together. Without us perhaps they would never have spoken to each other. And every day, under our Maple tree, we find pieces of dry bread and nuts.

乌鸦，布尔盖罗尼和枫树

自从我们在这里筑了巢后，一切都改变了。

以前这一切都是不可能的：他们不愿意和我们生活在一起。为了让我们离开，布尔盖罗尼家尝试了各种方法，他们甚至相信可以用那些愚蠢的毛绒玩具来吓唬我们。

但是，自从我们在他们的阳台的树上建了巢，我们便成为了焦点。

同时，楼上的卡斯托迪家和塞雷蒂家，照顾着我们孩子的成长。

我们知道他们在关注我们，有时，在我们进食的时候，他们会悄悄拨开窗帘凝视我们，只是为了不打扰到我们。

现在，他们已经习惯了我们的存在。我们成为将这栋楼里的三个家庭聚在一起的话题。

没有我们，也许他们永远也不会互相聊天吧！

每一天，在我们的枫树下，都会发现面包和坚果。

The Blackbird, the Evergreen Oak tree and the Longonis

The Longonis, so the next-door neighbours say, never go out. Or maybe they go out at night, using the service lifts. Someone swore they saw her, one cold winter in the street market in via Volturno, going to buy bird cages. Someone else (I think it was Nardozzo's wife) talked about him, kind and shy, at old Parotto's funeral. Some people say they have a house full of birds, that they catch them when they arrive on the big Evergreen Oak tree on the balcony. Some say that they eat them. But there are those who swear by their goodness and instead talk about the old Blackbird that the Longonis looked after. He arrived in a sorry state and today he's become a part of the family. He knows everything. But sitting on the highest branch day after day he sings his tunes, hiding his secret.

乌鸫，冬青栎和隆戈尼

隆戈尼一家，据他们的邻居说，从不下楼，也不出门。

或者他们只在晚上出来，用大楼的服务电梯。

有人发誓说见过女主人，是在一个寒冷的冬天，在沃土诺街的市场里，打算买几个鸟笼。

还有人，应该是那多佐的夫人，说起过这家的男主人。说他是一个和蔼但难以捉摸的人，曾参加过帕洛图老人的葬礼。

也有人说，他们家有个鸟房，有好多鸟，是当它们飞到他家阳台上的大冬青栎时被抓住的，之后便被养着。

但还有人信誓旦旦地说隆戈尼家一直照顾着一只老乌鸫。当初那只乌鸫受伤了来到他家阳台，如今已经成为他们家中一员。

它什么都知道。

它在最高的枝头哼着调，一天天，藏着它的秘密。

The Falcon, the Broom bushes and Brambilla

It turns gliding behind the spire and swoops like lightning on balconies in flower. It feeds on sparrows, mice; when things are bad it eats worms. Then it goes back up to the roof among the Broom bushes; looking down at the Park and the city. No one has even seen it up close, except Brambilla who goes up there every month to clean the big terrace on the roof. Perhaps there is complicity between the two. Maybe respect; or mutual fear. But everyone in the area knows the flight of the Falcon in the sky. Clean sinuous lines, then a sudden change of direction and acceleration as it darts between the towers and skyscrapers. A raptor must be frightening. The austere solitude of a Master of the sky.

猎鹰，金雀花丛和布莱姆比拉

它飞旋着绕过塔尖，如同闪电般穿梭于开满鲜花的阳台。

它捕食麻雀和老鼠，如果运气不好，也吃蠕虫。随后，它回到屋顶上金雀花丛的高处，仔细打量公园和城市。

没有人近距离见过它，除了每月打扫天台的布莱姆比拉。

也许，他们之间有着什么牵连。抑或尊敬，抑或互相恐惧。

但是，这里每个人都知道，在这个区域的天空中，有猎鹰的轨迹。

流畅地在空中盘旋，然后突然转弯，加速，冲向摩天大楼间。

猛禽定是令人生惧的。

空中霸主，冷峻而又孤傲。

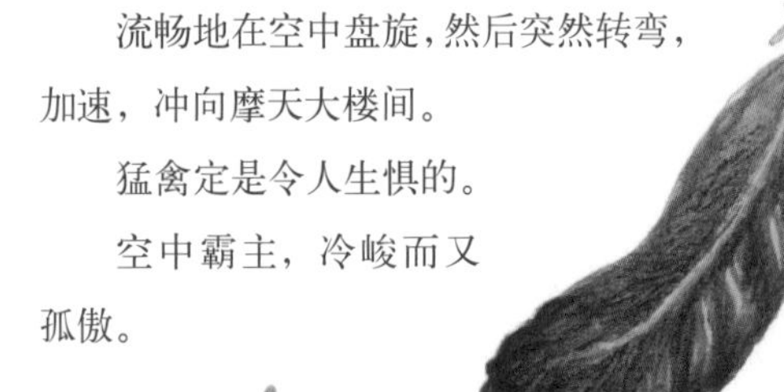

The Flying Gardeners

Ugo's always chewing on a bit of gum. Lino wears a bandanna, and Fausto's got dreadlocks. Once a year, roughly, they fly around the Forest. They hang from the edge of the roof on ropes and drop down jumping between the balconies, to prune the dead branches, cut away the untidy leaves, check the health of the trees. They studied arboricuture and then – who would have thought it? – they learned to be Climbers. And there they are now, gardeners floating in the Milan sky, appearing and disappearing in the frame of the big windows; a smile and off they go, down along the trunks with horizontal moves and vertical lines. Arboriculturists and climbers, only they have the consciousness of the richness of life in the Forest, because through looking after the trees they have also learned to look at the lives of people, in a sequence of intimate snapshots: the furniture, the disorder, paintings and dishes become clues to a life, the life of a couple, a family. Clues towards a collective comic of stories that the flying detectives collect, watching from outside through the leaves – the anger, the calm, the boredom, the sex, the solitude of vertical humanity.
Plant psychoanalysis.

飞翔的园丁

乌戈总是大嚼着口香糖。利诺扎着大花头巾。福斯托钟爱雷鬼造型。

大约每四个月，他们就绕着垂直森林飞翔。

他们挂着绳子从屋顶的边缘缓缓下降，在阳台之间穿梭，修剪枯枝，清理杂乱的叶子，检查树木的健康状况。

他们研究植物学。然后，谁曾想到？他们还学会了攀岩！

他们是漂浮在米兰空中的园丁，时而出现，又时而消失在大窗户的窗框间。一个微笑便转身，顺着树干横向或纵向移动。

他们有植物学家和攀岩者才有的，对于树木的丰富知识。

通过照顾树木，他们也学会了仔细观察人们的生活，一系列家庭的温馨的瞬间：陈设的家具、凌乱感，都是生活的痕迹，夫妻生活，家庭生活。

这些就像在飞翔中收集而成的生活漫画特辑，他们在外部，透过树叶观察着垂直人性的平静、嘈杂、性，以及孤独。

植物心理学。

The Swift, the Jasmine and Carletti the accountant

Poor Carletti the accountant, he just couldn't get over his mourning. The vegetable garden under the big maple tree had become a jungle; and he... well he never left the house. It was Rovetti's daughters who worked the miracle, along with their girlfriends on the thirteenth floor.
They went to find him a few days before the Summer Solstice and, quite unrepentant, they invited him to a balcony concert on the family terrace. He arrived wearing a blue shirt and flared trousers, and embarrassed and chatty, he never stopped smiling all evening.
Who knows if it's true what they say about the Rovetti girls, that they're strange and beautiful and love birds. It seems that once they found a Swift on the terrace. They knew it had come from far away and that it could not fly off again on its own. So they put it on the outside edge of the big balcony, beyond the Jasmine leaves and pushed it out into space. Sometimes, but only sometimes, you just need a little push.

雨燕，茉莉花和会计师卡尔莱蒂

会计师卡尔莱蒂迟迟无法从悲伤中恢复过来。

枫树下的蔬菜园已经杂草丛生，而他……一直地呆在家中。

但这一切改变了，是因为罗维蒂的女儿们和她们家在十三楼的伙伴们。

在夏至的前几天，她们去找他，邀请他参加露台上举办的家庭音乐会。

他来了，穿着一件天蓝色衬衫和喇叭裤，有点尴尬却十分健谈。一整晚，他都笑容满面。

谁知道关于罗维蒂家朋友的事情是真是假？据说，她们古怪、美丽，还十分喜欢鸟。

貌似她们有一次在露台上发现一只雨燕。它远道而来，无法独自起飞。

她们将它放在靠近露台边缘茉莉花树的叶子上，轻轻把它推向空中。

有时，但只是有时，轻轻一推，就足亦。

Bozzolo brothers' seeds

"It must have been that bitch in the other tower... that one who never stops watching us. Or the doorkeeper, with his snotty attitude. Or who knows, maybe some damn tourist, with his cretinous bird-watchers telescope. Me, I told you we shouldn't let it grow so much, that if it went over the edge of the balcony there'd be trouble, that at least we should have cut it a bit. But you, you're so damn pig-headed. And irresponsible, too. That's you all over. Now what do we do?
I told him I didn't know anything, that maybe it was the gardeners, or the seeds flew here from another balcony, and that in any case we'd get rid of them straightaway. But them, they didn't seem convinced. And threatening as well. And the next-door neighbours saw them come in, in uniform, and who knows what they thought... Well, shit, deep down they were laughing too. Maybe it would be enough to offer them some leaves, and a bit of a smile, what do you reckon? Anyway, it's all your fault. You said, a hundred meters up, it wouldn't grow. Well... now it's a forest... What the hell do we do?" "I know: let's switch to hortensias!"

波佐罗兄弟的种子

“一定是对面楼里那个坏女人……她一直在看着我们。也有可能是那个看门人，他总是态度傲慢。谁知道呢？或者是那些该死的游客，他们总是拿着望远镜朝我们这儿看鸟。我之前就和你说过，不要让它们长成这样，如果它们长到阳台外面很麻烦，至少要去剪一下。但你，怎么那么猪脑子，而且，不负责任。都是你的错！现在，我们该怎么办？

我和他说过，我什么都不知道。或许是园丁，或是从其他阳台掉下来的种子，但是我们会马上把它们处理掉的。但他们似乎不太相信，还威吓我。邻居们看到那些穿制服的人进来了，天知道他们心里在想什么？妈的，他们肯定在心里笑坏了。或许给他们一些叶子，再配以微笑，会有用吗？你觉得呢？不管怎么说都是你的错，你不是保证说，在百米高空，它不会长出来的么。看，已经变成一片树林了，我们到底该怎么办？”

“我知道，我们把它换成绣球花吧！”

VERTICAL FOREST IN FIGURES *

1.7 kms (approx.) of overall linear development of plant pots

2 hectares of forest (the equivalent of two soccer fields)

3 levels of protection provided by the anchor systems

4 - 6 annual pruning operations scheduled

6 - 13 kN/m³, soil density: dry (min)/saturated (max)

9 m (approx.) maximum tree height

10 m (approx.) maximum internal span of the structure

33 species of evergreen plants

50 - 110 cm, depth of plant pots

59 species of plants useful for birds

60 species of trees and shrubs

60 - 500 m², range of surface areas of the apartments

62 species of plants attractive to butterflies

65 species of plants adapted and suitable for insect populations

66 species of plants useful for pollinator insects

78 m (18 floors) the height of the Confalonieri tower

100 cm, the thickness of the soil substrate

94 different plant species

111.15 m (26 floors), the height of the De Castillia tower

131 apartments

190 km/h, wind speed in the plant test in the wind tunnel

300 kg, design weight of the largest tree (6 m), without turf/soil at the moment of planting

325 cm total balcony depth

480 human beings

600 kg, design weight of the largest tree (6 m), without turf/soil during its lifespan

711 trees

820 kg, weight of the largest tree including turf/soil at the moment of planting

1,600 birds and insects (approx.)

5,000 shrubs (approx.)

8,900 m^2, the approximate overall balcony surface area

15,000 perennials and drooping plants (approx.)

19,825 kg/year of CO_2 absorption (estimated quantity)

20,000 plants of different kinds (approx.)

40,000 m^2, built surface area (approx.)

100,000 m^2, expansion area of single-family homes (equivalence in terms of urban density)

* the data refer to the Vertical Forest built in Milan Porta Nuova and the year 2014

数据化的垂直森林 ✱

1.7 km（大约）种植盆直线延展总周长

2 hm^2 森林（相当于两个足球场的面积）

3 个级别的保护，由锚固系统提供

4 ~ 6 年的修剪操作计划

6 ~ 13 kN/m^3，土壤密度：干（最小）/ 饱和（最大）

9 m（大约）树木的最大高度

10 m（大约）结构内部跨度最大值

33 种常绿植物

50 ~ 110 cm，种植盆深度

59 种宜于鸟的植物

60 种灌木和乔木

60 ~ 500 m^2 的公寓套内面积范围

62 种吸引蝴蝶的植物

65 种适合昆虫繁衍的植物

66 种宜于授粉昆虫的植物

78 m(18 层) 高的孔法龙尼埃里大楼

100 cm 的土壤基质厚度

94 种不同的植物

111.15 m(26 层) 高的德卡斯底里亚大楼

131 套公寓

190 km / h，植物在风洞测试中承受的风速

300 kg，乔木（6m）最大设计荷载，不包括种植所需的草皮和土壤重量

325 cm 的阳台进深

480 人

600 kg，乔木（6m），在整个生命周期内可达的最大设计荷载，不包括种植所需的草皮和土壤重量

711 棵树

820 kg，乔木（6m）最大设计荷载，包括种植时所需的草皮和土壤重量

1 600 只鸟和昆虫（大约）

5 000 株灌木（大约）

8 900 m^2 阳台表面积总值（大约）

15 000 棵多年生植物和悬垂植物（大约）

19 825 kg/ 年的二氧化碳被吸收（预计量）

20 000 棵不同的植物（大约）

40 000 m^2 的表面积（大约）

100 000 m^2 的单户住宅城市密度水平铺开占用的面积

* 数据参考 2014 年在米兰新门建成的垂直森林

ILLUSTRATED DICTIONARY OF THE VERTICAL FOREST IN 100 ITEMS

edited by Guido Musante

编辑：圭多·穆桑特

100 个垂直森林元素的插图字典

Ambasz, Emilio An Argentine architect, Ambasz (Resistencia, Chaco, June 13th 1943) is one of the forerunners of so-called "green architecture" and in some ways of the vision encapsulated in the Vertical Forest. Ambasz's design approach, described by himself as "green on grey", involves the use of large roof gardens and green roofs: elements of a neo-natural landscape as opposed to the built urban landscape. His architectural manifesto is ACROS (Asian Cross-Road over the Sea), a multifunctional complex located in the Tenjin district of Fukuoka, Japan. Opened in April 1995, the ACROS is typified by its south facade, punctuated by a series of terraced gardens and covered by dense vegetation. Initially, the complex housed 76 different plant species, giving a total of approximately 35,000 plants; with the passing of time, birds and the elements have helped to increase its biodiversity, introducing seeds of new species into the micro-ecosystem and bringing to 10 the number of varietes, as well as a total of 50,000 plants pesent today.

>> see Basic radicality

安巴斯，埃米利奥 埃米利奥·安巴斯（1943年6月13日出生于查科省，雷西斯滕西亚）是一位阿根廷建筑师，他是“绿色建筑”和垂直森林愿景的先驱之一。安巴斯的设计方案被他自己描述为“灰色之上的绿色”。其方案涉及大型空中花园和绿色屋顶的应用，使新式的自然景观元素与建成的城市景观形成对比。他的建筑宣言是ACROS（海洋上的亚洲十字路）——位于日本福冈市天神区的一座多功能综合楼。该项目于1995年4月启动，一系列植被稠密的台阶花园构建出建筑的南立面，这是ACROS的精华。最初，这栋综合楼里有76种植物，总共有约35 000株植物。寒来暑往，在鸟类和天气的作用下，新的物种的种子被带到了微生态系统中，从而又新增120个新品种。如今，已共计有约5万株植物。

>> 参见：激进的原型

Anchor system The presence in the Vertical Forest of trees planted at great heights involves the study of specific security solutions for the inhabitants and users of the public spaces below. The first Vertical Forest, created in Milan's Porta Nuova, is provided with an anchor system of trees based on three levels of protection:

· a temporary security device: all medium and large trees are anchored to a horizontal frame of tubular elements fixed to the bottom of the pots by means of security straps holding the turf, which prevent the tilting of the trunk away from its turf and potential exit from the pot;

· a base fixing device: all medium and large trees are fixed via three elastic straps to an aerial steel restraining cable anchored to the floor of the terrace above, which is designed to prevent the falling of the tree in unexpected extreme conditions, such as the breaking of the trunk, and which is able to adapt to the growth of the plant over time;

· a redundant safety device: trees subjected to the worst environmental conditions - such as a wind speed higher than that allowed for in the design - are equipped with a steel basket designed to bind the turf to the concrete structure. The system consists of two cross-members, bound to the upper elements of the vertical frames, which physically prevent the escape of the root-ball and turf from the pot.

>> see wind tunnel, structure, Pots

锚固系统 特为垂直森林中高空生长的植物而设，是为保障居民和公共空间的使用者安全而做的专项研究成果。第一座垂直森林建于米兰新门，其树木锚固系统具有三重安全防护措施。

· 临时安全装置：所有中型和大型植物的树根土球被固定在种植盆中由管柱组成的水平框架上，种植盆上覆盖草皮，草皮上面又用安全皮带来固定。这样可以防止树干倾斜或树木从草皮下，甚至从种植盆中翻出来。

· 基本固定装置：通过3根弹性安全皮带，所有中型和大型植物被固定在一个悬空的拉伸钢索上。钢索一端接在阳台地坪上，一端接在上层阳台板上。这样可以预防植物在某些极端条件下倾覆或枝杆断裂。安全皮带可以根据树木的生长情况定期调整。

· 保险的安全装置：树木一旦遭遇极端恶劣的环境状况，比如风速超出了设计所允许的速度，那么还有一个钢制篮将树根球固定在混凝土结构上作为保护。这个体系由两根交叉的横梁组成，并束缚在垂直框架的上部，这消除了树根土球和草皮从盆子里倾覆出的可能性。

>> 参见：风洞，结构，种植盆

(Anti-) anticity The level of proximity and complication between the spaces that the presence of nature helps to create within the Vertical Forest can generate forms of forced interaction and effects of open socialization - the exchange of knowledge, unexpected examples of altruism - that turn out in many cases to be the opposite of those usually created in the sprawl of suburban landscapes, or the so-called "anticity" (cf. Stefano Boeri, Anticittà, Laterza, Bari, 2011).

While the model of the anticity is composed of a myriad small closed systems, one set alongside another yet not communicating, the model of the city condensed in the Vertical Forest is based on a number of different systems, set in highly dense conditions which interact continuously and with a high rate of variation

>> see Anti-sprawl device, Boundaries

反－反城市化 因有了自然的存在，各个空间的关系变得更为亲密和复杂，这有助于在垂直森林中构建强制的互动，并形成开放式的社会化。知识的交流及许多未曾预料到的利他主义的案例，在很多情况下被证明，在郊区蔓延中是无法实现的。我们称之为“反城市化”（博埃里，反城市化，2011）。

反城市模型，是由无数个小型封闭系统组成的。一个系统虽然与另外一个非常接近，却不存在交流。但浓缩在垂直森林里的城市模型却基于许多不同体系之上，这些体系间存在高密度且有连续性的互动，并一直处于较高变化率的环境中。

>> 参见：反城市蔓延的设备，界限

Anti-sprawl device Right from its earliest conception, the Vertical Forest has attempted to give substance to the hypothesis that it should be possible to create houses in dense and central city areas with the same relationship of intimacy and proximity to greenery that characterizes life in suburban areas and in the country. Developed on this basis, the Vertical Forest design aims to transpose onto a central metropolitan area lifestyles and ways of life that are typically found in individual houses and cottages, widely distributed throughout the outskirts of Italian and European cities or in more general terms, the areas of the so-called 'sprawl'. Assessed in terms of urban density, each tower of the Vertical Forest is equivalent to an expansion area of single-family homes of about 50,000 square metres of surface area. The height development of the buildings was determined by the need to leave as large as possible an area of land free to be used as a park and public space. As a 21st century urbanization experiment, the Vertical Forest is intended as an anti-sprawl solution, that is, able to induce the abandonment of forms of urban sprawl that have defined the twentieth century and which have

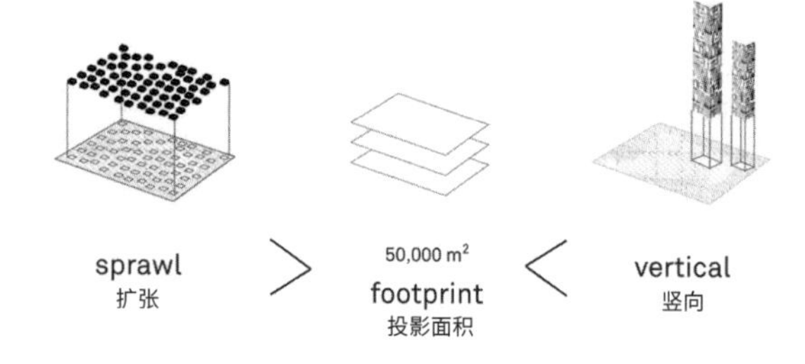

produced cities that often lack quality and identity.

>> see (Anti-) anticity, Boundaries, Densification, Urban traffic

反城市蔓延的设备 在最早期的概念中，垂直森林意图成为一个实体，即使在高密度的城市里，甚至市区中心，也能够犹如在农村和郊区一样，与自然植物亲密接近。以此概念为基础，垂直森林的设计进一步发展，旨在将大都市中心区的生活方式与郊区独户别墅的典型生活方式相嫁接。这些独户别墅的生活区，广泛地分布于意大利及其他欧洲城市郊区，也就是所谓的城市“蔓延”区域。从城市密度的评估中可知，每栋垂直森林相当于 50 000m² 的独户别墅用地。建筑的建设高度，应尽可能地多留出空地来建设公园或作为其他公共空间以满足需要。作为 21 世纪城市化的一个尝试，垂直森林将作为一个反城市蔓延的设备，引导人们去放弃 20 世纪形成的城市蔓延区，因为在这些城市蔓延区上建设的社区，通常欠缺品质与特性设计方案。

>> 参见：反－反城市化，边界，致密化，城市交通

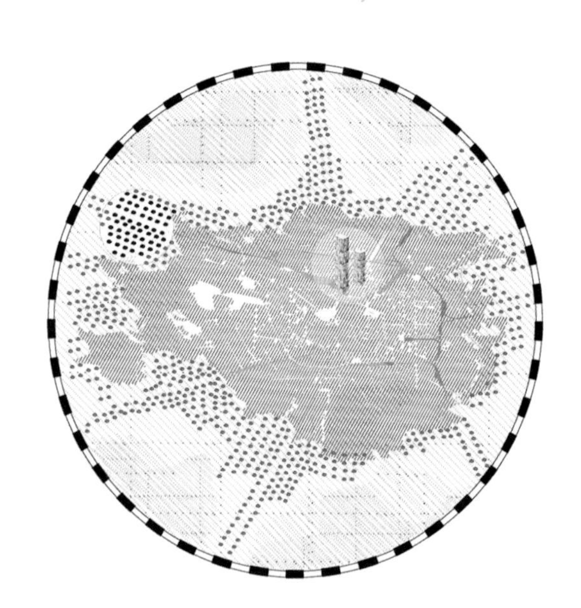

Balconies From an architectural standpoint, the balconies are the most important element of the Vertical Forest. Balconies integrate pots that house plant species, they provide the ideal environment for the formation of microhabitats, they constitute the extension space for the lived internal areas and meetings points between human and animal and plant life. If metaphorically equating the Vertical Forest building to a large tree, the balconies can be considered to constitute the branches. In the first Vertical Forest created in Milan the balconies have three basic characteristics, all interconnected: the location for plants pots set along the edges (integral with the parapets), the significant overhanging over the edge of the façade, and a mutual staggering or offsetting. Interestingly, this last feature wasn't part of the early versions of the project, which allowed for balconies on each floor that extended in continuous strips along the whole width of the façade.

The initial solution was then replaced by one based on staggered balconies, each different from another in linear development along the facades and variously distributed on the different floors. This version, which eventually proved to be the definitive one resulted in a series of different intermediate levels from balcony to balcony - single, double or even triple – which was useful for the grafting of plants with heights up to nine metres. The most obvious consequences of this design choice can be seen not only in the highly dynamic design, as perceived from the façades, but also in the influence established in the perception of the 'boundary' between one balcony and another.

The second important feature of the balconies of the Milan Porta Nuova Vertical Forest is their major overhang over the front edge of the façade. In their final configuration, they all extend out for a distance of 3 metres and 25 centimeters. This solution has allowed an expansion of the inhabited spaces in the open air and at the same time the creation of plant pots with a greater depth (up to 110 centimeters). The overall surface of the balconies is approximately 8,900 square meters.

>> see Colours, Boundaries, Three-dimensional façade, Big tree, Structures, Soil, Pots

阳台 从建筑学的角度看，阳台是垂直森林中最重要的元素。阳台与种植盆一体化，种上植物的阳台为微生态提供了理想的生成环境。阳台还构建出室内生活的延伸空间，提供人类、动物与植物接触的区域。如果将垂直森林比喻为大树，那么阳台则可以理解为树枝。第一栋垂直森林建于米兰，它的阳台具有三个基本特性并互为关联：种植盆沿阳台外缘布置（与护墙成为一体），显著地突出于建筑物的立面，且彼此交错。有趣的是，垂直森林最终呈现的这个特色外观与项目最初的构想并不完全相同，在原先设想中，每一层的阳台是像带子一样围绕着建筑体，在整个立面宽度方向上延展的。

最初的方案被阳台错落的方案所取代。每个阳台形态各异，沿着立面呈线性排列，分布于不同的楼层。这个方案之所以具有决定性的意义，在于它能在竖向的阳台与阳台之间形成一系列不同的高度空间，有单层高、双层高、甚至三层高，这对于种植 9m 高的大树实在太有帮助了。选择这个设计的意义，不仅仅呈现在其非常动感的立面形态，还体现在阳台之间的“边界”感。

米兰新门的垂直森林阳台，沿着建筑立面向外悬挑是其第二个重要特性。在最终的外形中，这些阳台均向外悬挑出 3.25m。这个方案，使得内部居住空间得以向室外空间有一定的扩展，且种植盆的深度可达到 1.1m。阳台的总面积大约为 8 900m^2。

>> 参见：色彩、界限、三维立面、大树、结构、土壤、种植盆

Basic Radicality he Vertical Forest harks back to design images and principles related to radical ecological thinking from the Sixties on, (among them those by authors such as Gianni Pettena, Frederich Undertwasser, the Site group and Emilio Ambasz). Like the radical non-anthropocentric visions, and as opposed to the demands of the Modern Movement, in the Vertical Forest, green is not "domesticated" by architecture, but rather an element that "relativizes" or adds a sense of proportion to the project, or even - in metaphorical terms - "corrodes" it. However, unlike what was frequently seen in radical architecture, in this case the process of conceptual reversal does not take on extreme forms. In the Vertical Forest architecture is certainly present in second place to plant life, but this is in a non-emphasized, even basic style.

>> see Ambasz, mute architecture, Beuys, biodiversity, Hundertwasser, sustainability

激进的原型 垂直森林可以追溯回 20 世纪 60 年代激进生态思潮下的设计图像和设计原则，其中部分由詹尼·佩戴那、百水先生、场地组合和安巴斯等创作。垂直森林与激进的反人类中心主义的愿景相似，也与现代建筑运动的要求相反。在垂直森林，绿色并非指建筑中"家养的"植物，而是在项目中占相当比例的元素，它起到绿"化"建筑的作用，甚至于可以用"侵蚀"这个词来比喻它。然而，不同于我们常见的激进派建筑，在这个设计方案中，概念的逆转并没有以一种极端的形式出现。对于垂直森林而言，建筑相较于植物生命居于次位，这并不会显得很突兀，反而是本质的回归。

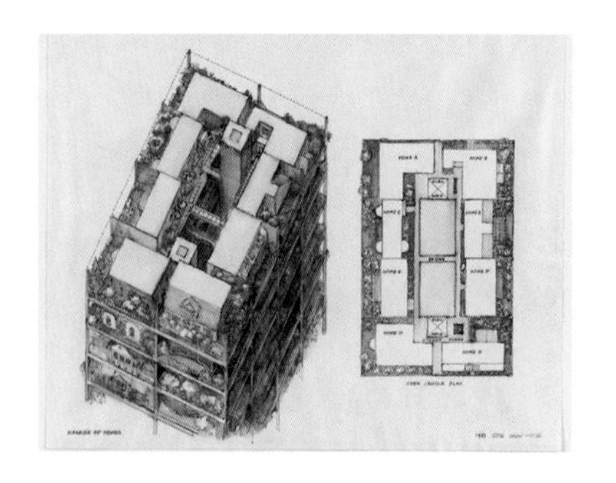

>> 参见：安巴斯，沉默建筑，博伊斯，生物多样性，百水先生，可持续性

Beuys, Joseph Joseph Beuys (Krefeld, 12th May 1921 – Düsseldorf, 23rd January 1986) was a German painter, sculptor and artist. His work anticipates some of the themes architecturally interpreted by the Vertical Forest. In Beuys' vision, the act of planting trees becomes a collective and founding ritual, capable of evoking the deeper meanings of the relationship between man and nature. In 1972 in the small village of Bolognano (Pescara) in Italy, Beuys started a series of initiatives in which artistic, political and ecological themes were intertwined, and this is how the Foundation Institute for Revival of Agriculture (1976) and the Piantagione Paradise (Paradise Plantation) came about, with the planting of 7,000 trees for the restoration of biodiversity (1982). All of these ventures were part of the Difesa della natura operation which continued with various projects until 1985. One of these in 1982 was 7,000 eichen (7,000 oaks), created to mark the seventh edition of the Documenta exhibition in the German town of Kassel. The work consists of a large triangle, set in front of the Federiciano Museum and which comprises 7,000 basalt stones, each of which can be 'adopted' by a potential buyer. The progressive replacement of the stones by trees came to mean a gentle collective alchemy, which would establish a deep symbiosis between trees, people and cities.

>> see biodiversity, humans

博伊斯，约瑟夫 约瑟夫·博伊斯（1921年5月21日生于克雷菲尔德，1986年1月23日卒于杜塞尔多夫）是德国画家、雕塑家和艺术家。他的作品中预见了一些垂直森林中诠释的建筑主题。在博伊斯看来，植树这一行为是一个集体的奠基仪式，可以唤醒人与自然关系中更深层的内涵。1972年，在意大利佩斯卡拉的小村庄博洛尼亚诺，博伊斯开始了一系列的创作，其中错综复杂地融入了艺术、政治和生态主题。这也就是农业复兴基金学院（1976）和种植乐园的由来。其中，他为生物多样性的复兴种植了7 000棵树（1982）。他的这些尝试都是自然防御行动的一部分，并同各种各样的其他项目一直开展到了1985年。1982年的项目之一——7 000橡树，是第七届卡塞尔文献展的标志。这个作品呈一个巨大的三角堆，矗立在腓特烈博物馆之前，它由7 000块玄武岩石头组成，每一块石头都可以被一位潜在的买家以一颗橡树作为交换来“领养”。这种用树换石头的方式，蕴含着温柔的集体主义的魔力，建立起植物、人与城市之间深入的共生互利的关系。

>> 参见：生物多样性，人类

Big Tree In metaphorical and conceptual terms, the Vertical Forest can be compared to a big tree, where the balconies are the branches, all the plant species are the leaves, the central body of the building is the trunk and the roots are water supply systems. On the one hand the metaphor of the big inhabited tree sums up the character of the architecture of biodiversity in which the Vertical forest is the “bearer”, while on the hand it refers to an imagery that has been widely looked at in the literature of fantasy and fable (think for example of the Peter Pan of James Matthew Barrie, the Baron in the Trees by Italo Calvino or Midworld by Alan Dean Foster).

>> see balconies, colors, the Baron in the Trees

大树 以比喻和概念性的词汇来解释，垂直森林好比一颗大树，阳台是树枝，所有品种的植物都是树叶，建筑的中心体是树干，树根则好比供水系统。一方面，垂直森林作为供人类居住的大树“载体”，概括了生物多样性建筑的特性；另一方面，它又展示出一个幻想文学与寓言文学中经常出现的景象（可以回想下詹姆斯·马修·贝瑞的《彼得潘》，卡尔维诺的《树上的男爵》以及阿兰·迪恩·福斯特的《中世界》）。

>> 参见：阳台，色彩，《树上的男爵》

Biodiversity (graft of) The Vertical Forest is a grafting of biodiversity onto the built city environment. Its presence creates an urban ecosystem

- rocky and mixed - in which different types of vegetation create a habitat that can be colonized by birds and insects, and that can become a focal centre of attraction and a symbol of spontaneous re-colonization of the city by plant and animal life. The various types of green in the Vertical Forest - tree, shrub, suspended lawns, vertical green connections between the arboreal centres - give life to an environment able to support rupiculous (rock or wall-dwelling) bird species (like redstarts or kestrels) and/or other species (such as tits, goldfinches and turtledoves), terrestrial vertebrates (bats) and insects (such as Diptera, spiders, butterflies and certain beetles).

The creation of a series of Vertical forests in an urban setting can generate a network of environmental corridors able to connect with the habitats of urban parks, green spaces, avenues and gardens, the natural vegetation of the city. The terraces and roofs of the first Vertical Forest created in Milan are dedicated microhabitats and the site for a program of targeted interventions aimed at the supporting of animal life, the installation of artificial nests, the planting of host plants and the placing of feeders suitable for providing nourishment for certain species (while avoiding attracting unwanted species). All these activities are subject to monitoring and data collection for scientific observations. Among the main objectives of the study group for biodiversity active in the Milano Porta Nuova Vertical Forest is a series of activities in support of meta-populations of pollinators such as bumblebees and solitary bees: species that from the Fifties onwards in Italy have been subject to a drastic decline. The program has been followed through the installation of artificial nests and the grafting of plant species suitable for supporting the habitat of insects and which have flowering periods that last as long as possible throughout the productive season.

>> see urban forest, continuous cities, mineral cities, ladybugs/ ladybirds, densification, changing landmarks, sustainability, plant species

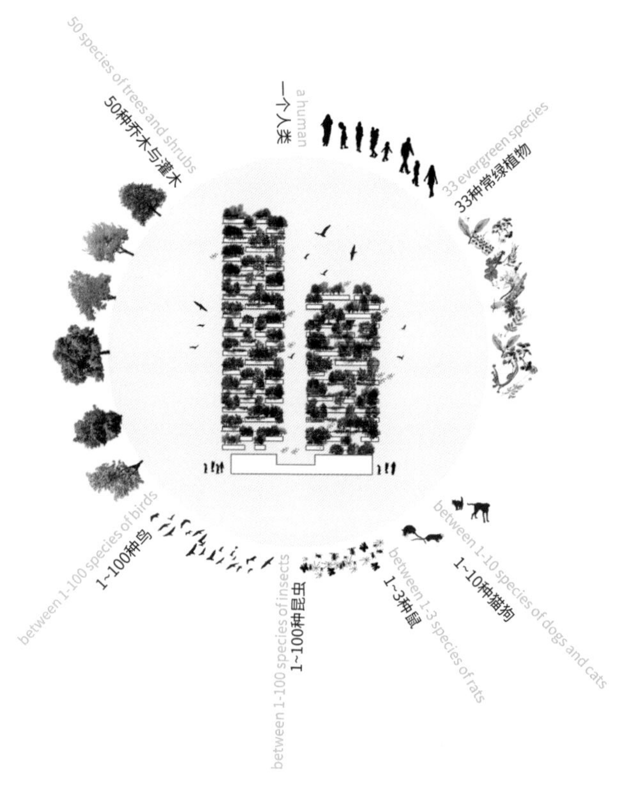

生物多样性（移植） 垂直森林，它将生物多样性嫁接到了城市建成环境中。它的出现创造了一个城市生态系统：石质的、混杂的，在它之中不同类型的植物又为鸟类和昆虫创造出一个栖息地。它是一个吸引力的中心，是城市中动植物重新自发地移居与再繁衍的聚焦之地。垂直森林里各种类型的绿色——乔木、灌木、空中的草坪、树栖中心间的垂直绿色区域，都提供了可以让鸟类（比如红尾鸲、小鹰、山雀、金翅雀和斑鸠）、陆栖脊椎动物（蝙蝠）和昆虫（如双翅目、蜘蛛、蝴蝶和某些甲壳虫）栖居的一个生存环境。

在城市环境中建设一系列的垂直森林，可以形成一个环境走廊的网络，它将城市公园、绿化区、林荫道、花园，以及城市的自然植被联系在一起。米兰的首个垂直森林的露台和屋顶是订制的微生境，旨在为动物的生存提供有计划和有目标的支持。其中，人造鸟巢的安置、可寄宿的植物种植以及饲养器的布置，都可以为某些目标物种提供营养（同时，也避免吸引一些不想要的物种）。所有这些举措都有助于科学观察、监控和数据采集。研究小组在米兰垂直森林的主要目的之一，就在于通过观察一系列生物活动，来支持诸如大黄蜂和独居蜂等传粉者聚居种群的生存和发展。自 20 世纪 50 年代以来，意大利境内这些物种突然急剧减少。该项目从始至终，都进行着人造鸟巢的安置，移植适合昆虫作为栖息地的植物物种，选择的这些植物有漫长的开花时间，并与昆虫繁殖季节重合。

>> 参见：城市森林，连绵城市，矿物城市，瓢虫／瓢甲科，致密化，变化的地标，可持续性，植物品种

Blackbird [*Turdus merula*] A bird of the Turdidae family found all over Europe (with the exception of northern Scandinavia), Asia and northwest Africa. In northern areas during the winter it migrates to warmer climes, while in temperate zones it can be found all year round. Its favoured habitat is the forest, but it can adapt to live in other environments, such as orchards and vineyards or even urban areas in close contact with man. It has the ability to easily learn any melody and then repeat it ad nauseam. It sometimes happens that two distant blackbirds produce alternate sounds, singing different phrases without one interrupting the other.

画眉【乌鸫】 属于鸫科，在欧洲地区（除了北斯堪的纳维亚外）、亚洲和非洲西北部都很常见。北方一到冬天，它们就迁徙到更温暖的地方，而在温带地区，一年四季都可以看到它们的踪迹。这些鸟最喜欢的栖息地当属森林，但是它们可以适应其他环境，比如果园和葡萄园，甚至与人类关系密切的城市区。它们具有快速学会任何旋律的能力，然后没完没了地重复。有时候，两只相隔较远的画眉鸟会交替地哼唱，它们唱着不同的乐曲，却丝毫不会干扰到对方。

blackcap [*Sylvia Atricapilla*] A sedentary passerine bird of the Sylviidae family, characterized by a strong body and grey plumage. The populations found in the northern and central European areas winter in southern Europe and North Africa, where there are also local populations. In recent years a number of specimens from Central Europe have developed the habit of wintering in the gardens of southern England and Scandinavia: it is assumed that the availability of food and the possibility of avoiding migration over the Alps have prevailed bearing in mind the presence of a less than optimal climate. It prefers sady woodlands with ground cover for nesting.

黑顶林莺【黑顶林莺】 莺科的一种定栖的雀形目鸟，以强健的体格和灰色的翅膀为特征。这种鸟

群可在欧洲北部和中部找到，而冬季它们则迁至欧洲南部和北非地区，这些地区也有其本地种群。近几年，人们从一些来自中欧地区的样本中发现，它们也在英国南部和斯堪的纳维亚的花园越冬。据此推测，食物供应能力和避免飞越阿尔卑斯山迁徙的可能性都说明气候条件并不乐观。这些鸟类更喜欢在阴凉的林地用地被植物筑巢。

processes with the potential to be repeated at other times and in other places.

书 《一座垂直的森林》这本书是一本解释性的科学手册，对垂直森林的特性和意义进行描述、呈现与分析，以 2014 年落成于米兰的第一座垂直森林的原型建筑为开篇。这本手册创作的理由是将垂直森林作为一项多学科实验的焦点。这些实验活跃在不同的层面上（建筑学、城市学、社会学、生物学、行为学等）。其中一个实验的目的在于为未来制订好步骤、预测结果，找到潜在的可实施性条件及拓展文化视野。此外，这本书还是一本将虚构的故事与真实的想法以一种创造性的叙述方式表现出来的小说。总之，这本书所呈现的垂直森林，不仅仅是一个完成的建筑物，更确切地说，它是一个开放的过程集，它完全可以在不同的时间和地点中重演。

Book The A Vertical Forest book was originally created as an explanatory scientific manual offering a description, presentation and analysis of the characteristics and meanings of the Vertical Forest, starting from the first example built and opened in Milan in 2014. The justification for the manual comes from the idea of the Vertical Forest as a multidisciplinary experiment, active on several different levels (architectural, urban, social, biological, ethological, ...): an experiment from which it is hoped to derive procedures, results, applicative possibilities and cultural horizons for the future. The book is also a novel that brings together invented stories and real ideas in an imaginative narrative dimension. Overall, the book presents the Vertical Forest not as a finished architectural object but rather as an open set of

Boundaries In the Vertical Forest, green is a common good which goes beyond the boundaries established by individual properties. This assumption brings together different aspects of the project. On the one hand, the vertical development makes the flora and fauna of the Vertical Forest an integral part of the urban landscape, which everyone can enjoy. On the other, the connections in environmental terms that develop between the habitat of the Vertical Forest and the surrounding urban green areas help to diminish the sense of limits between "private building" and "public space". This diminishing of the sense of limits can also be found within the Vertical Forest. In the Milano Porta Nuova Vertical Forest the interplay of irregularity between balconies, necessary for the upward growth of the larger trees, has produced two types of potential "invasion" of the space of others: that of more developed plants, and that of the looks or glances of the inhabitants. The combination of these two aspects has blurred the perception of 'limit' between one balcony and another, and between one private property and another. The resulting relativization of the sense of intimacy and the change in the dynamics of socialization of the residents constitutes a unique and experimental element of the project. Within the Vertical Forest the traditional ideas and sense of limits in terms of 20th century models of living have been completely outdated.

>> see anti-anticity, balconies, anti-sprawl device

界限 在垂直森林，绿化是一项凌驾于私有物业边界之上的公共利益。该举措综合考虑到了项目的方方面面。一方面，垂直发展可以使垂直森林中的动植物群受大众喜爱，且成为城市景观中必不可少的部分；另一方面，垂直森林里的自然栖息地与周围城市绿化区之间的关联，有助于减少“私人住宅”与“公共空间”之间的临界。在垂直森林当中，也可以找到这种临界的消逝。在米兰新门的垂直森林中，阳台之间不规则的交错，为较大植物向上生长提供了必要的空间，由此，形成了两种潜在 “侵入”其他空间的情形：侵入生长茂盛的植物，侵入或观望其他居民的生活。这两种“侵入”的结合，模糊了一个阳台与另外一个阳台的“界限”，以及一个私人领地与另外一个私人领地的“界限”。将私密感与居民动态社交行为的变化进行关联，构成了这个独特的实验项目。在垂直森林里，20 世纪生活模式的传统观念以及所谓的界限感完全都过时了。

>> 参见：反　反城市化，阳台，反城市蔓延的设备

california lilac [*Ceanothus spp*] Shrub or small tree of the Rhamnaceae family, native to North America and especially California. Generally no more than 0.5-3 metres high, it can grow on dry, sunny hills or coastal areas. It is often grown as an ornamental garden plant. Native Americans used the dried leaves as a herbal tea, which was then adopted by settlers as a substitute for black tea.

加州紫丁香 【美洲茶】 鼠李科的一种灌木或小型乔木，原产于北美洲，尤其是加利福尼亚州。通常高度为 0.5～3m，它可以在干燥、阳光充足的山上或沿海地区生长。经常被作为花园的观赏植物。美洲的原住民们经常把晾晒干的紫丁香叶作为花草茶，之后的殖民者则把它作为黑茶的替代品。

Celentano, Adriano Celentano (Milan, January 6th 1938) is an Italian singer, dancer, actor, director, record producer, editor and television presenter. In the lyrics of several of his songs he has explored various issues related to the relationship between man-city and nature. These include a song from 1972, a symbolical anticipation of the poetry of the Vertical Forest: A tree 30 floors high (Ouch. I can not breathe / I feel / that I'm choking a little/ I feel my breath going down,/ it goes down and doesn't come back up / I only see that / something is / emerging ... / maybe it's a tree / yes it's a tree / thirty floors high.)

阿德里亚诺·切伦塔诺 （1938 年 1 月 6 日出生于米兰）是一位意大利歌手、舞蹈家、演员、导演、音乐制作人、编辑和电视主持人。在他的部分歌曲中，探讨了关于人、城市与自然之间关系的各种各样的问题。其中一首 1972 年的歌曲——《30 层高的树屋》——对垂直森林进行了诗意般地展望：喔，我简直无法呼吸了，我感觉有点窒息，呼吸越来越困难，越来越困难，无法恢复，我只看到某些东西在显现…… 或许是一棵树，哦是的，那是一棵 30 层楼高的树。

chaffinc [Fringilla Coelebs] A small passerine bird of the Fringillidi (finch) family widespread throughout Europe, North Africa and Asia. It prefers to live in woods, among scattered trees and bushes, in hedgerows, fields, orchards and wherever there is vegetation. During the winter it often frequents cultivated fields or the outskirts of cities, where it is easier to find food. It is fond of cold climates and is often found in mountain areas. It has a particular song, similar to the sound of a doorbell.

花鸡【苍头燕雀】 一种体型很小的鹰科类雀形目鸟。这种鸟广泛分布在欧洲、北非和亚洲地区，喜欢住在树林里，在散植的乔木、灌木丛间，灌木篱墙中，田野里、果园以及其他任何有植物的地方。在冬季，花鸡经常会去耕地和市郊，因为那里比较容易寻觅到食物。这种鸟喜欢寒冷的气候，经常出没于山区。它们歌喉独特，与门铃的声音有些相似。

changing landmark The towers of the Vertical Forest not only offer their inhabitants an extraordinary perspective from within the apartments. Cyclically changing their skin according to the diversity of the plants and their disposition with respect to the sun's axis, the Vertical Forest also offers a changing landscape to city dwellers. The colour elements vary significantly throughout the year: in spring above all the pastel shades stand out, while in autumn and at the end of the growing season the warm colours become more pronounced. The changes in the image of the Vertical Forest are not only due to the change of colours, but also the textures of the solids and voids formed by branches and leaves. The colour display of the first Vertical Forest was defined through a careful selection of the greenery to be planted, primarily horizontal in the shape of shrubs and perennial plants of which there are 100 different species, many of which are native.

>> see colours, biodiversity, plant species

winter
冬

summer
夏

autumn
秋

spring
春

变化的地标 垂直森林这栋大楼不仅让其居民从室内就可以观赏到非凡的美景，大楼的外观也会因为植物多样性以及太阳光轴的变化，产生出周期性的变化。此外，垂直森林还为城市居民提供了一个千变万化的风景。颜色元素一年到头都会有非常明显的变化。春季，一片清淡柔和的色彩，而到秋季，生长季节结束之时，暖色调则突显出来。垂直森林形象上的变化，不仅仅归因于颜色的变化，枝干和花叶所构成的实体与空隙的纹理也是原因之一。首栋垂直森林的色彩，基于精挑细选的植物配置，包括了 100 多种不同的灌木和多年生植物，其中很多植物是本土的。

>> 参见：色彩，生物多样性，植物品种

collared dove [Streptopelia decaocto] A bird of the Columbidae family native to Asia, but widely found in Europe since the twentieth century. It prefered habitat is arid and semi-desert areas with few trees, but in recent years it has become increasingly present in populated areas, especially in urban parks with evergreen trees which provide good shelter.

灰斑鸠【灰斑鸠】一种原产于亚洲的鸠鸽科鸟，但自 20 世纪后就广泛分布在欧洲地区。它们喜欢居住在树木较少的干燥的半荒漠地区，然而近几年来，

在居民区，尤其是在种着常青树、可以为它们提供阴凉的城市公园，灰斑鸠变得越来越多。

Colours If observed as a highly complex chromatic system, the Vertical Forest can be described as having two primary components: the floral and faunal elements and the mineral body of the buildings. In the first Vertical Forest the principal colour component of the building is that of the outer cladding, consisting of large porcelain tile sheets in matte finish gunmetal gray. The colour of the tiles almost perfectly matches the glass and frames of the large floor to ceiling windows, which alternate on the facades with a variable pattern. The colour equivalence between tiles/windows emphasizes the monolithic appearance of the two towers and renders them comparable to a pair of large tree trunks, reinforcing the architecture-tree metaphor.

In terms of perception, the 'deaf' or 'mute' colour impact of the construction element of the Vertical Forest helps to bring out the vibrant volumes and changing colours of the leaves, branches and shrubs. The design also allows for the use of white inserts as a device for varying the composition. On all the balcony balustrades porcelain sheets of the same size as the cladding tiles have been inserted but in gloss finish screen printed white glass: a solution that creates a highly dynamic pattern on the facades with a musical rhythm. Strips of the same colour and material have also been attached to the overhanging sections of the balconies, providing a colour match with the lower surfaces. This solution emphasizes the already significant staggering of the balconies - especially when viewed from below - and their assimilation as the metaphorical branches of the big tree.

>> see big tree, balconies, changing landmarks, materials

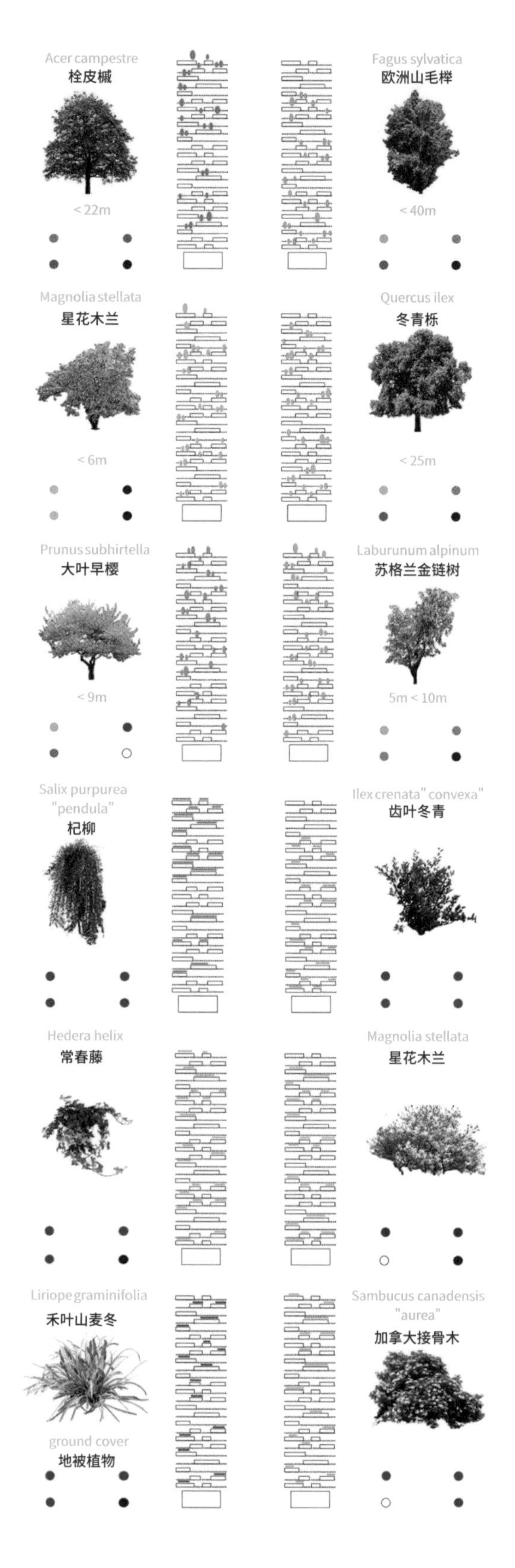

色彩 若把垂直森林视作一个高度复杂的色彩体系，则可将其描述成是由两种主要成分构成的：动植物元素以及建筑物的矿体。在首个垂直森林中，建筑物的主要色彩来自其外墙：暗灰色的喷砂陶瓷板。陶瓷板的色泽几乎可以完美地与落地窗的玻璃和框架搭配，通过多变的图案组合，装饰在立面上。陶瓷板与窗户协调的色彩强化了两栋大楼外观的统一性，色彩的渲染使得它们犹如一对双生大树，体现出建筑和树木间的隐喻关系。

从感知出发，垂直森林的建筑元素带来的“聋”或“哑”的色彩效果，可以衬托出树叶、树枝和灌木的生机勃勃与多变的颜色。设计还植入了白色来作为变奏。所有的阳台栏杆上，都贴上和外墙瓷砖一样大小的板材，嵌入抛光丝印的白色玻璃。这是一种通过音乐韵律，创造高度动态图案的设计。同样颜色和材质的条板也被应用到阳台的外悬部分，与地面的颜色搭配非常好。这种设计突显了阳台，尤其是当你从下往上看的时候，更会感到阳台犹如从大树上舒展开的枝干一般。

>> 参见：大树，阳台，变化的地标，材料

common broom [Cytisus scoparius] Small herbaceous shrub belonging to the Fabaceae family. It is present throughout Western Europe and parts of Scandinavia and grows at different altitudes, from sea level to about 1,400 metres above sea level. It ranges from 60 centimetres to 1.3 metres in height and is woody at the base, with bushy and wintering buds of fragrant at complexity, the Vertical Forest is true consists golden yellow. During cold periods the grassy sections dry out and only the woody and underground parts remain alive. The fruit is a flattened black legume.

紫雀花【金雀花】 小型草本灌木，属于豆科植物。在整个西欧和斯勘的纳维亚半岛的部分地区可见，它可以在海平面到海拔 1 400m 的区间中生长。高度为 60cm 到 1.3m，根部颜色为木色。它有灌木和越冬植物的芽儿和芳香，在垂直森林中呈现金黄色。在寒冷的季节，叶子都枯萎了，只有枝干和地下的根系还生机勃勃。

common hawthorn [*Crataegus monogyna*] Shrub or small tree belonging to the Rosaceae family, widespread throughout Europe, North Africa, West Asia and North America. It has a large number of spiny branches and can reach heights of between 50 centimetres and 6 metres. Its natural habitat is woods and bushy areas, mainly in calcareous soils and it grows at altitudes of between 0 and 1,500 metres above sea level.

山里红【山楂】 属于蔷薇科的灌木或小树，广泛分布于欧洲和北非、西亚和北美。它有很多带刺的树枝，高度在 50cm 到 6m 不等。其天然产地为树林和灌木地区，适生于石灰土质，可以在海平面到海拔 1500m 的区间内种植。

construction timing Construction of the first Vertical Forest in Milan began in Autumn 2009 and finished in Autumn 2014 (the official opening was held on October 10th 2014).

施工期 米兰的首个垂直森林在 2009 年秋季动工，并于 2014 年秋季竣工（2014 年 10 月 10 日正式开盘）。

continuous city The Vertical Forest is a grafting of biodiversity onto the Continuous City. In many parts of the planet – Europe in particular but also parts of North America and Asia - the development that has taken place in recent decades has created urban areas similar to one huge polycentric city, which has absorbed a huge number of small and medium-sized cities. The Continuous City has profoundly altered biological balances and reduced biodiversity within itself, incorporating natural and agricultural areas, accelerating the reduction and extinction of plant and animal species, or even in some cases forcing certain species, forcibly removed from their natural habitat, to invade urban areas that are foreign to them. The Vertical Forest was created from the observation of these phenomena, and at the same time from the realization that such abnormal situations could not be addressed by a mere act of returning to the past or contraction of the cities within their traditional boundaries, but within the conditions imposed by the Continuous City.

>> see Biodiversity, Densification, Urban sensor, Sustainability

连绵城市 垂直森林将生物多样性移植到了连绵城市中。在地球上的很多地方，特别是欧洲、北美和亚洲的部分地区，经过近几十年的发展，形成了巨大的、类似于吸收了许多中小城市的多中心城市。连绵城市严重地破坏了生态平衡，减少了生物多样性，通过合并自然和农业区域，加速了动植物的减少和灭绝，甚至于某些物种被迫从他们天然的栖息地转移到完全陌生的城市地区去。垂直森林就是基于这些现象的观察而创建的。设计者意识到，这种不正常的情况不能够通过仅仅还原到过去，或者收缩城市规模来解决，而应当根据连绵城市的现状来加以解决。

>> 参见：生物多样性，致密化，城市传感器，可持续性

Critical Opinions The first example of the built Vertical Forest revealed the project's immense capacity to be a powerful generator of critical opinions in a variety of forms, contents, and vehicles of expression. The style and tone of recorded comments was also very different (among others for example see: Vittorio Gregotti, Le ipocrisie verdi delle archistar, Corriere della sera, 18th February 2011; Mario Botta, Viva la città moderna ma Milano non è Abu Dhabi, La Repubblica, 23rd November 2007; Antonio Cipriani, Politica, media, città: vivere in un format, www.globalist.it, 28th February 2014).

>> see documentaries, echoes

批判性观点 第一座垂直森林的建造，已经诱发人们通过各种形式、内容和表达媒介来传递他们对于这个项目的意见。评论的风格和语气也千差万别。（请参考如：维托里奥·格力高帝，建筑明星虚伪的绿色，晚邮报，2011 年 2 月 18 日；马里奥·博塔，万岁，现代城市，是米兰不是阿布达比，共和报，2007 年 11 月 23 日；安东尼奥·希普利亚尼，政治，媒体，城市：生活在格式中。www.globalist.it，2014 年 2 月 28 日）。

>> 参见：纪录片、反响

Cultural cell The Vertical Forest is a system that increases urban ecological culture. The service "cell" in charge of guaranteeing the functionality and maintenance of the greenery over time can become a reference point of information for urban para-ecosystems both for neighbouring schools and for the city. Each "vertical green maintenance cell" can be used for the gathering and spreading of information that will be useful for assessing its ecological functioning (conducting census of species or colonization, fluctuations etc.) amassing an amount of knowledge that will grow and evolve together with the Vertical Forest.

>> see maintenance, urban sensor

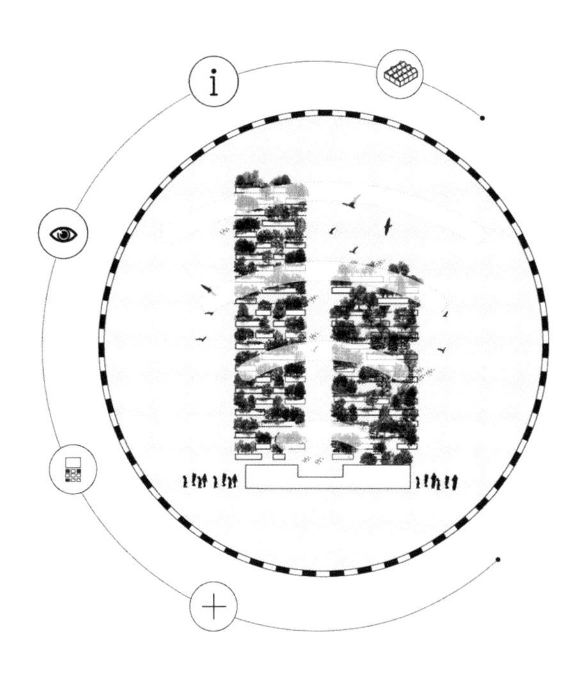

文化细胞 垂直森林是一个增强都市生态文化的系统。这个“细胞” 的职责在于：确保绿色植物的长期功效，提供养护，并成为周围学校以及这个城市生态系统的信息参照标杆。每一个“垂直绿化养护细胞”都可以用于收集并传播有用信息，这些信息有益于评估其生态功能（统计物种数量、移殖以及波动等情况），并随着垂直森林一起发展、积累、演变为知识库。

>> 参见：养护、城市传感器

Demineralization The Vertical Forest is a demineralization project for urban surfaces. It reduces the heat emanating from public spaces produced by the reflection of the sun on the mineral facades of tower buildings clad with curtain wall facades of steel and glass, and the project is also opposed to their pattern of energy consumption.

>> see mineral city, energy device

去矿化 垂直森林是一个对城市表面去矿化的工程。它减少了由于太阳辐射到覆盖着钢铁和玻璃的矿物建筑表皮而对公共区域所产生的热量。这还是一个降低能源消耗的工程典范。

>> 参见：矿物城市，能源设备

Densification The Vertical Forest is a model of vertical densification of greenery and the built environment constructed within the city that acts in conjunction with the policies of reforestation and naturalization of the great urban and metropolitan frontiers. The synergy between these two tremendous devices for environmental survival can facilitate the reconstruction of the relationship between nature and cities in the context and area of the contemporary city. The Vertical Forest acts through completely changing the principle of closed green spaces (gardens, parks) that is typical and widespread in the fabric of many (especially European) cities and offers a system for vertical greenery that is as adaptable as it is well-structured.

>> see anti-anticity, biodiversity, continuous city, anti-sprawl, urban sensor

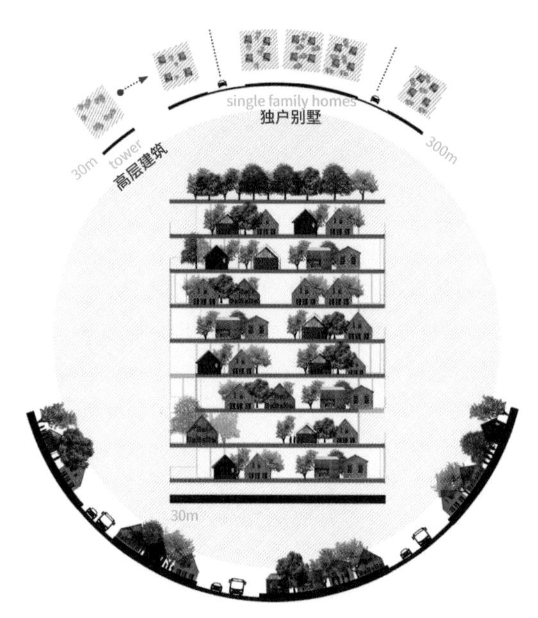

致密化 垂直森林是一个垂直绿化和建成环境的致密化模型。它也是与在都市边界鼓励重新造林与自然归化的政策密切相关的模型。垂直绿化与建成环境的融合，在协同拯救环境的同时，促进重建起当代城市与自然的和谐关系。垂直森林的建设将改变许多城市、建筑、绿地空间（公园和花园）的建造原则，并为发展垂直绿化提供一个合适且较好的结构体系。

>> 参见：反－反城市，生物多样性，连绵城市，反城市蔓延的设备，城市传感器

Documentaries An experimental habitat with a large number of unknown features, the Vertical Forest is particularly suited as the subject of scientific investigations and documentary narratives in ways that have few parallels in architecture and is more comparable with what usually occurs in a naturalistic context. Its unique features are ideal for leading into a development of innovative narrative techniques. Already during the construction phase the first Vertical Forest in Milan has been the subject of various studies and documentaries. In March 2014 a programme produced by the British BBC One Planet broadcast team saw the adoption of television shooting systems that had never been used before. Positioned on the roof of facing Garibaldi tower, the BBC cameras captured a series of sequential images of the Vertical Forest over more than 18 months. In order to show the process by which the building site became an architecture / living habitat in its entirety, special technology was used that combined three different filmmaking modes (time-lapses, hyperlapses, tracking-lapses). The use of this combined technique, extended over a long period of time, resulted in a highly involving and dynamic filmed narrative, capable of generating a strong sense of perceptual participation in the process of the Vertical Forest development. The resulting movie allows the viewer, for example, to appear on the site, to fly to the foot of the building, follow a tree while it is being lifted up to the top of a tower and observe the plants as they grow and change in appearance throughout the seasons.

>> see cultural cell, echoes, critical opinions

纪录片 垂直森林是具有许多未知特性的实验性栖息地项目，它特别适合作为科学调查和纪实叙述等主题影片的题材，记述建筑的发展与自然环境的发展，并进行平行比较。其独特性可以为叙述手法带来创新与发展。米兰的首个垂直森林还在施工的时候，就已经成为各类研究与纪录片的主题。2014年3月，英国BBC制作了节目《一个星球》，使用了以前从未用过的电视拍摄系统。BBC的摄像机被安置在加里波第塔的屋顶，连续超过18个月，相继为垂直森林捕捉到一系列的影像。为了全面展示垂直森林是如何从建筑工地变成一栋建筑和一个生存环境的过程，拍摄人员使用了将三个不同制片模式（延时拍摄、快速移动和追踪拍摄）结合起来的特技。这种拍摄技术延续了很长一段时间，最终做成了一部颇受称赞的动态电影叙述片，让人们可以很直观地感受垂直森林开发的过程，仿佛置身其中。例如，制作出的电影可以让观看者如临现场，或飞到建筑物的下方，或跟随树木被吊升到大楼顶部，甚至看着植物生长，观察它们的四季变化。

>> 参见：文化细胞、反响、批判性观点

Downy oak [Quercus pubescens] The downy oak is the most commonly-found species of oak in Italy. Able to adapt to both arid and relatively cold climates, it has great strength and plasticity, thanks to the tremendous vitality of the stump: qualities that allow it to acclimatise easily to different spatial conditions. Unlike other species of oak, in the winter the dry leaves remain attached to the branches. In ancient times specimens of downy oak were planted along the boundaries of a property, allowing later generations to reconstruct and date the latter.

毛栎【柔毛栎】 毛栎是意大利最常见的树种之一，可以同时适应干旱和较寒冷的气候。它有很大的强度和可塑性，这要归功于树桩的巨大活力：它的特性可以适应不同环境中的各种空间情况。不像其他橡树，到了冬季，它的干叶仍然附着在树枝上。在古代，柔毛栎沿着领地的边界栽种，以此作为后辈们重建的依据。

Echoes Due to its unique experimental nature, the first Vertical Forest, created in Milan, has aroused widespread media attention, echoes and interest from

the press and critics around the world, not just from the architectural sector. Among the first awards for the project was the 2014 International Highrise Award and the award as best 2015 European architecture by the Council on Tall Buildings and Urban Habitat, promoted by the Illinois Institute of Technology of Chicago.

>> see documentaries, critical opinions

反响 由于其独特的实验性质，第一座垂直森林在米兰一经建成，就引起了来自世界各地新闻和评论界的广泛关注、反响和兴趣，不仅仅是在建筑领域。在项目获得第一个大奖——2014 国际高层建筑大奖后，又获得了由芝加哥伊利诺伊理工大学推动的世界高层建筑与都市人居学会授予的 2015 欧洲最佳高

Financial Times - leading daily review / UK

Daily newspaper

Date 8-9-11-2011

Page 1, 12

House & Home

Ed Hammond
Size matters: measuring the pay gap in sq ft
Perspective Page 5

Dressed to impress
The Frizzing art of super-prime selling UK property Page 4

Lady Judge
City views almost as impressive as her CV
At home Page 2

Ebenezer Howard's first 'Garden City' in Letchworth was so influential that in 1907 Lenin stayed there

The age of flower towers

Architects are tackling the problems of the concrete jungle with ambitious schemes using green technology to grow forests in the sky. By Christopher Woodward

层建筑奖。

>> 参见：纪录片，批评性观点

Energy device The Vertical Forest can be designed to house wind turbines on the roof, photovoltaic panels for the production of electricity on the roof and facades, and solar panels for the production of thermal energy. Irrigation of the plants can be organized largely through the filtration and purification of waste water produced from within the building. The shield formed by plants located along the perimeter of each floor results in a significant reduction of energy consumption produced by artificial air-conditioning machinery. In these terms, the Vertical Forest is an alternative to large residential towers with glass curtain walls which require considerable en-

ergy to compensate for the considerable accumulation of heat in the summer and the lack of protection from the cold during the winter, and the resultant low energy efficiency even when amply equipped with renewable energy supply systems.
>> see mineral city, demineralization, sustainability, laboratory roof

能源设备 垂直森林可以装配屋顶的风力发电机、屋顶以及立面上的光伏板，以及产生热能的太阳能板。大楼里产生的污水在经过过滤和净化后，可用于灌溉植物。由每一层环绕种植的植物所形成的遮蔽和保护，可以显著地减少空调系统所产生的能源消耗。综合这些因素，较之那些有着玻璃幕墙、需要大量能源来进行夏季降温和冬季避寒的居民楼，垂直森林系统是个不错的选择。因为传统玻璃幕墙建筑即使充分配备了可再生能源供应系统，其能源的利用效率仍然不高。
>> 参见：矿物城市，去矿化，可持续性，实验室屋顶

Eurasian Wren [*Troglodytes*] A passerine bird of the Troglodytidae family, it is common in Europe, North Africa, Asia and North America. It has a round body, is just 10 centimetres long and moves in an agile and responsive manner. Sedentary, living mainly in humid places with an abundance of bushes, it prefers to move on the ground, inspecting everything that interests it. In winter it can be seen on the ground close to residential areas, while in summer it prefers mountain areas.

山蛔蛔【鹪鹩】 鹪鹩科的一种雀形目鸟，在欧洲、北非、亚洲和北美都比较常见。圆体型，只有 10cm 长。行动敏捷，反应快，习久坐。主要生活在具有大量灌木的潮湿地区。喜欢在地上走，观察一切感兴趣的事物。冬季，在居民区附近的地面上也可以看到它的踪迹，而夏季，它们则喜欢在山区活动。

European Wild Pear [*Pyrus pyraster*] A plant belonging to the Rosaceae family, it has a native area stretching from Central and Western Europe to the Caucasus. It needs the cool-temperate climes of its native areas where it grows both on the plain and in the mountains, sometimes even reaching 1,400 metres above sea level further south. It can develop as a simple shrub, 3-4 metres high, but in ideal conditions it can develop into a tree of up to 20 metres. It is the ancestor of the common cultivated pear, from which it is often difficult to distinguish.

梨【欧洲野生梨】 一种属于蔷薇科的植物，其原产地从欧洲中西部到高加索地区。喜低温气候区，可以在平原和山区生长，在南部地区有时甚至可以在海拔 1 400m 的地方生存。它可以长成 3～4m 高的灌木，不过，在条件较好的地区，它可以长到 20m。一般很难辨认它和普通梨树，因为它是普通梨树的始祖。

Evergreen oak [Quercus ilex] A plant belonging to the Fagaceae (beech and oak) family, commonly found in the countries of the Mediterranean basin, especially in the west, Algeria, Morocco, the Iberian peninsula and in Mediterranean France and Italy. It can produce single species woods and copses of large dimensions. When it grows in rocky environments it can develop a bushy appearance. Usually evergreen, it has a rarely straight trunk, more commonly single or divided at the base with a height up to 20-25 metres.

常青树【冬青栎】 属于壳斗（山毛榉和橡树）科的一种植物，通常在地中海盆地（尤其是西部，阿尔及利亚）、摩洛哥、伊比利亚半岛以及地中海的法国和意大利等国家比较常见。它可以产生单一品种的树木和较大尺寸的小灌木丛。当它在多岩石环境下生长时，可以长得非常浓密。通常，它是一年四季长青的，树干非常直，大部分是直生的或者在根部分支，其高度可达 20～25m。

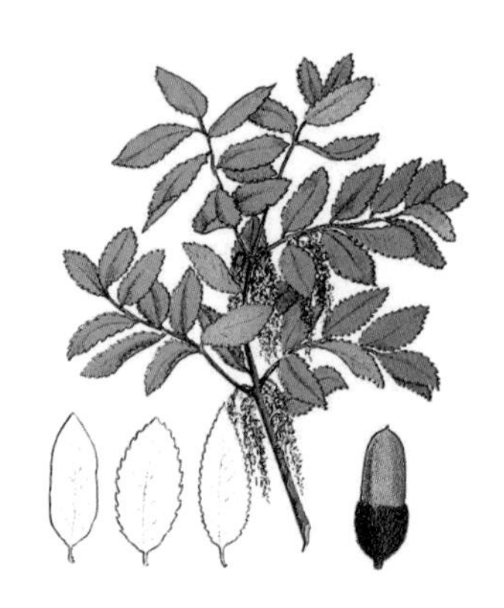

Expo 2015 The first Milan Vertical Forest is the prototype of a strategy of biodiversity acupuncture that large cities today are increasingly called upon to plan. In this sense, the project is presented as the symbol of the challenge for biodiversity that the Milan Expo 2015 threw down to the inhabitants of the planet.

>> see biodiversity

2015 年米兰世博会 米兰的首个垂直森林是大城市寻求生物多样性策略的一个原型。在这个意义上，对于这个星球上的居民而言，这个项目成为 2015 年米兰世博会展挑战生物多样性的象征。

>> 参见：生物多样性

Falcon [*Falco*] A breed of bird of prey of the Falconidae family, found worldwide. Equipped with a small head and a highly aerodynamic body shape, they can weigh up to 2 kgs. Very rapid in flight and when nose diving, they are capable of capturing live prey both in the air and on the ground. They do not build nests, but lay and hatch their eggs in old nests left by other birds in hollow trees, on rocky crags, or in depressions they dig in the ground.

猎鹰【隼】 隼科中食肉鸟的一种，遍布于世界各地。头小，身体呈现流线形，重量可达到 2kg。飞行速度非常快，在头部朝下俯冲时，可以捕捉空中和地面的活猎物。它们无需自己筑巢，而是在其他鸟类废弃的旧巢孵小鹰。这些巢有的筑在空心树上，有的位于岩石峭壁上，有的位于它们在地面挖的洼坑内。

Feral pigeon [*Columba livia*] A semi-wild pigeon widespread throughout Europe, North Africa and the Middle East which is strong and fast in flight. It is 30-35 cm long and has a wingspan of 62-68 cms and its flight capability is significant: under optimum atmospheric conditions it can cover 800 kms at an average of 70 km/h. It is granivorous and prefers squares and parks in big cities as a habitat where it easily coexists with the presence of man. It commonly nests on buildings, especially under eaves or in gutters.

野鸽【原鸽】 遍布欧洲、北非和中东地区的半野生鸽子，体格强健，飞行速度非常快。长度 30～35cm，翼展有 62～68cm，其飞行能力非常强，

在适宜的大气环境下，它可以以 70km/h 的速度飞行 800km。它属于食谷类鸟，喜欢将大城市的广场和公园作为栖息地，在这里它们可以与人类共存。通常，它们会在房屋上筑巢，尤其是屋檐或水沟旁。

Field maple [*Acer campestre*] Small tree widespread in Europe and Asia. In Italy it is very common in temperate deciduous forests, from sea level up to the start of beech forests. It is frequently used as an ornamental tree and hedge, because of its effectiveness in consolidating and resisting landslides, and its leaves can be used as fodder. Normally it is about 7-12 metres high, but it can also grow up to 20 metres. Folklore attributed magical properties to it against witches, bats and misfortune.

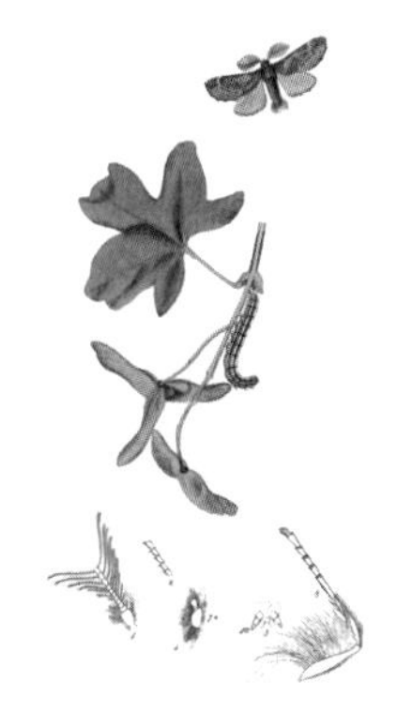

篱槭【栓皮槭】 产于欧洲和亚洲的小树。在意大利的温带落叶林，这种树木十分常见，从海平面到山毛榉森林分布的高度，这种树都适于生长。因

为它在巩固土层和抵抗滑坡方面功效非常显著，所以这种树经常被用作观赏树木和树篱。此外，其树叶也可以做饲料。正常情况下，这种树 7～12m 高，但是，它也可以长到 20m 的高度。根据民间传说，野生枫树在防御巫婆、蝙蝠和灾难方面有着神奇的魔力。

Goldenrain Tree [*Koelreuteria paniculata*] Deciduous tree of the Sapindaceae family, native to China, Korea, Japan and now widely cultivated in southern Europe as an ornamental plant. It does not grow to great heights (usually no more than 12 metres) and has pinnate leaves that turn yellow in autumn. The flowers appear about halfway through the summer and are small, yellow and gathered in pyramidal spikes. The fruits are yellow-brown heart-shaped bladder-like pods about 4-5 cm long.

木栾【栾树】 无患子科的一种阔叶树，原产于中国、韩国和日本，如今已作为观赏植物广泛种植于欧洲南部。这种树一般不会长得特别高（通常不超过 12m），其羽状叶子在秋天会变成黄色。夏天可以看到花。栾花黄色，是呈三角锥状的穗状花序。果实呈黄色或棕色，心形囊状，就像 4～5cm 长的豆荚。

Goldfinch [*Carduelis Carduel*i] Migratory bird belonging to the Fringillidi (finch) family, easily recognizable by the vertical red mask on the face and the wide yellow bar on the wing. Widespread in Europe, North Africa and Western Asia, it inhabits open plains or partially wooded areas. During the 19th century it was introduced into many areas of the world including Australia and New Zealand. It is often bred in captivity for its distinctive appearance and pleasant singing. Its name comes from the thistle plant of whose seeds it is particularly fond. It was adopted as an iconographic symbol in many representations of ancient Greece and of Catholicism.

红额金翅雀【红额金翅雀】 属于雀科的候鸟，非常容易辨别。面部有红色面具，翅膀有宽大的黄色条状羽毛，广泛繁殖于欧洲、北非和亚洲西部。喜欢栖息在辽阔的平原或者局部树木繁多的地方。在 19 世纪，它被引进到世界上很多地区，包括澳大利亚和新西兰。由于其外表特别，歌喉动人，因而经常被饲养赏玩。在古埃及和天主教中还被当成一种图形化标识。

Great Spotted Woodpecker [*Dendrocopos*] A bird belonging to the woodpecker family, it is found throughout Europe, except in the northernmost regions and some islands. In Asia it is found more in Japan, China and western India. Very adaptable, it lives in both deciduous and coniferous woods, in trees in the countryside and also in urban parks. It can nest anywhere from valley floors to the higher reaches of forests, digging a nest in a wide range of types of tree, particularly large chestnuts, larches, poplars and cherry trees. In February-March it defends its territory in a lively way, with sharp calls and an energetic tapping.

Great Tit [*Parus Major*] A passerine bird of the Paridae family, it is found throughout Europe and North Africa, especially in the hills and plains or medium altitude mountains up to 1,800 metres. It lives in open woodland, the edges of forests, orchards, vineyards, gardens and urban parks. It adapts very well to the changes brought about by man to the environment: it accepts food offered in feeders and is one of the few species of birds found regularly in city centres. Wintering and migratory, in Italy it can be found throughout the year, especially in winter. It nests in the protected cavities of trees, walls and nesting boxes, building the nest with moss, hair and feathers.

赤鴷【大斑啄木鸟】 啄木鸟科的一种鸟。在除了最北边地区和一些小岛外，整个欧洲都可以看到它的踪迹。亚洲则在日本、中国和印度西部比较多见。它们的适应能力非常强，可以生活在落叶类树林和针叶类树林中，既可以在乡村，也可以在城市公园的树上栖息。它们在谷底或更高的森林里筑巢，各种不同的树都可以成为其栖居之地，尤其是较大的栗子树、落叶松、白杨树和樱桃树。2 月到 3 月，它们会通过尖叫和不停地啄木来保卫其领土。

灰山雀【大山雀】 山雀科的一种雀形目鸟，在欧洲、非洲北部，尤其在中等海拔到海拔 1 800m 的平原和山区比较多见，栖息于开阔的树林、森林、果园、葡萄园、花园和城市公园的边缘。它们对人类改造自然环境所带来的变化适应很快，接受喂食器提供的食物，也是城市中心常见鸟之一。它们是越冬类鸟，随着季节变化而迁徙。在意大利，一年四季都有，尤其是冬季。大山雀常在树洞、墙洞和巢箱里用苔藓、毛发和羽毛筑巢。

Greater Mouse Eared Bat [*Myotis Myotis*] A bat of the Vespertilionidae family, widespread in western and eastern Europe, Anatolia and the Middle East as far as Palestine. In Italy it is found everywhere on the mainland, but not in Sardinia. It lives in different types of habitats, including urban areas up to 2,000 metres above sea level and it can also withstand temperatures up to 45 °C while in summer it lives together in colonies of up to several thousand individuals in caves, mines, cellars and sometimes in tree holes and bat boxes. From September to April it hibernates in underground environments, with temperatures of between 2-12 °C and humidity up to 100%, where it forms colonies of up to 5,000 individuals.

大鼠耳蝠【大鼠耳蝠】 蝙蝠科的一种蝙蝠，广泛分布于东欧和西欧、安纳托利亚、中东以及巴勒斯坦。在意大利的大陆，不包括撒丁区，都有这种蝙蝠。它生长在各种类型的栖息地，包括城市区域和海平面高达 2 000m 的地方，可以承受温度高达 45℃的天气。在夏季，大鼠耳蝠生活在一起，在洞穴、矿井、地窖，有时在树洞和蝙蝠箱，几千只一起居住。从 9 月到次年 4 月，它们会选择在温度 2～12℃，湿度达到 100% 的在地下环境下过冬，多达 5 000 只蝙蝠一起聚居。

Greenfinch [*Carduelis Chloris*] A passerine bird of the Fringillidi (finch) family, it is widespread throughout Europe, North Africa and Asia Minor. It adapts easily to any habitat, although it prefers the zones above 1,000 metres a.s.l.. It can be found in wooded countryside, forests, orchards, but also gardens and public parks; it perches on trees and bushes and then jumps to the ground. It is one of the most common and widely found birds in Italy, where it lives all year and is most numerous in winter as a result of migration from Northern Europe. In recent years it has been threatened by trapping and poison administered to trees to control pests.

欧金翅雀【欧金翅雀】 雀科的一种雀形目鸟，广泛繁殖于欧洲、北非和小亚细亚地区。尽管更喜欢栖息在海拔 1 000 多米高的地区，但实际上它们对栖息地适应能力非常强，在树木繁盛的乡村、森林、果园、公园和花园都有。它们会栖息于乔木、灌木丛中，然后跳到地面。这是意大利最常见的一种鸟之一。在意大利全年都适宜这种鸟居住，尤其在冬季，由于北欧地区越冬鸟类的迁徙，意大利的金翅雀数量更多。近几年，为了防止树木被害虫毁坏，金翅雀的生存也受到了抓捕和毒害的威胁。

Guinigi Tower The Guinigi Tower in Lucca (Tuscany, Italy) is a historical case of "Vertical Forest before its time." Built in the late 14th century by the powerful Lucchese Guinigi family of traders and bankers, the 44.25 metres high stone and brick tower is one of the historic buildings of this Tuscan city. Among the many medieval towers in Lucca, the Guinigi tower is the only one that was not damaged or destroyed in the sixteenth century and is a highly recognizable urban landmark. On its top there is a small hanging roof garden, a symbol of rebirth which consists of seven oak trees, grafted onto a walled casement filled with earth. Exactly when the garden was created is not known but in an image of Lucca in the 15th century from the chronicles of Giovanni Sercambi, there is a tower crowned with trees; the oak trees present today on the tower were however surely replanted later.

>> see changing landmarks

圭尼基塔 意大利卢卡市的圭尼基塔，是垂直森林产生前的一个历史实例。圭尼基塔，由势力强大的圭尼基家族的贸易商和银行家们于 14 世纪创建。塔高 44.25m，由石头和砖块砌筑而成，是托斯卡纳市历史最悠久的建筑之一。在卢卡市的许多历史建筑中，圭尼基塔是唯一一座没有在 16 世纪被损坏或毁掉的建筑，它被公认为是城市地标。在其顶部，有一个小型的空中花园，墙体围合起的树池内移栽了 7 棵橡树，这 7 棵橡树被认作是“重生”的象征。空中花园建立的具体时间不得而知，但是，根据乔瓦尼 · 塞拉康比在 15 世纪《纪事报》中关于卢卡市的描述，当时就有一座顶部冠以树木的塔。如今这座塔上人们所看到的橡树，可以肯定是后人重新种植的。

>> 参见：变化的地标

Heights Upward development is a basic design characteristic of the Vertical Forest. The two residential towers that make up the first Milan Porta Nuova Vertical Forest have different heights of 111.15 metres, equivalent to 24 floors for the De Castillia Tower (formerly Tower E) and 78 metres, equivalent to 17 floors for the Confalonieri Tower (formerly Tower D).

>> see structure, Guinigi Tower

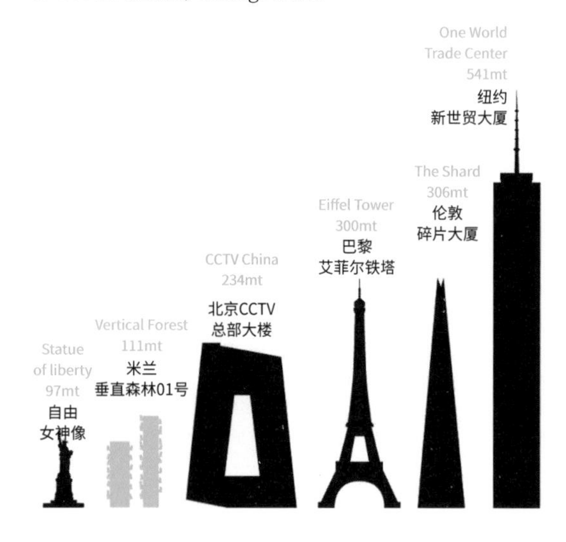

高度 向上发展，是垂直森林的一个基本设计特色。构成米兰首个垂直森林的两栋居民楼有不同的高度。德卡斯底里亚大楼（原称E楼），111.15m高，24层；孔法龙尼埃里大楼（原称D楼），78m高，17层。

>> 参见：结构，圭尼基塔

Hooded Crow [*Corvus cornix*] A passerine bird belonging to the Corvidae family, widespread in Eurasia and North Africa. Hooded crows are among the best adapted birds to urbanization: while maintaining their natural wildlife status they are accustomed to the presence of humans and often live in cities by finding food anywhere, even taking it from food put out for pets. Sometimes they create homes in trees surrounding houses and communicate daily with humans. They are not birds of prey but can be harmful to other birds, such as those feeding chicks and hatching eggs, and often cause damage to agriculture.

冠鸦【冠小嘴乌鸦】 一种乌鸦科的雀形目鸟，广泛分布于欧亚地区和北非。冠鸦是对都市化环境适应能力最强的鸟类，在维持其自然野生状态的同时，已经适应了与人类共存。它们乐于生活在城市中，因为便于寻找食物，甚至会抢夺宠物的食物。有时，它们会在房屋周围的乔木中筑巢，每天与人类交流。它们并不是猛禽，但是可能会对其他鸟造成伤害，比如那些鸡雏和种蛋。此外，它们还经常会给农业带来破坏。

House Martin [*Delichon urbicum*] A bird from the Hirundinidae family, it is found throughout Europe, Asia, north-western Siberia, Japan and north-western Africa. It lives up to an altitude of 2,000 metres and builds its nest under the eaves, canopies, walkways and spaces found in buildings or bridges. Since 2004 it is among those species at risk of extinction, and its nests are protected in several countries. In particular, changes in

building technologies are having a negative effect on the growth of the population of these birds: in actual fact nests do not adhere to the smooth facades and sealed surfaces that typify much contemporary architecture.

毛脚燕 【白腹毛脚燕】 属于燕科的一种鸟，常见于欧洲、亚洲、西伯利亚西北部、日本和非洲西北部地区。它们生活在海拔 2 000m 处，在屋檐、天棚、走道和建筑或桥梁的空间筑巢。自 2004 年来，毛脚燕就成为了一种濒危物种，它们的鸟巢在很多国家都是受到保护的。尤其是建筑技术的改变对这些鸟类的生长造成了很大的负面影响，因为它们的鸟巢无法建于现代建筑典型的光滑密闭的表面上。

Howard, Ebenezer Sir Ebenezer Howard (London, 29th January 1850 – Welwyn, 1st May 1928) was an English town planner and Esperanto-speaker. Inspired by the ideas of his fellow countrymen John Ruskin and William Morris, Howard devoted himself to the serious problem of overcrowding in cities and the consequent depopulation of the countryside as a result of the Industrial Revolution. In 1898 he wrote the essay A Peaceful Path to Real Reform, considered one of the most important theoretical texts on the concept of utopian cities in the late 19th century. In this work he describes the idea of the “garden city”, an urban area of well-established dimensions carefully designed to accommodate population movements through an efficient use of space and thus promoting a new balance between town and country. Howard does not describe the formal characteristics of the new city, but focuses exclusively on social and economic aspects, describing the specific functional destinations with great scientificrigour: dwellings, production establishments, and infrastructure and open spaces.
Howard’s theories did not find many direct applications: however among these the most important example is Letchworth Garden City, founded in 1903. His theories constituted an important reference point for the development of the Siedlung in Vienna in the early 20th century.

霍华德，埃比尼泽 埃比尼泽·霍华德爵士（1850 年 1 月 29 日生于伦敦，1928 年 1 月 5 日卒于韦林）是一位英国的城镇规划师和世界语演说者。受他的同胞约翰·罗斯金和威廉·莫里斯思想的启发，霍华德投身于工业革命所导致的城市拥挤和乡村人口减少的严重问题当中。在 1898 年，他撰写了《通往真正改革的和平之路》的文章，被认为是 19 世纪下半叶乌托邦概念中最重要的理论性文本。在这部著作中，他描述了“花园城市”的概念，即一个有着行之有效规模的城市区域，可以适应人口流动，通过有效地利用空间来发展新城镇之间的平衡。霍华德并没有描述新城的形式特征，而是完全注重于社会和经济方面，他描述了非常科学和严谨的特定功能目标：住宅，生产设施，基础设施和开放空间。霍华德的理论并没有找到很多直接的应用，相较之下，最重要的案例应是莱奇沃思花园城市，建设于 1903 年。他的理论对 20 世纪的维也纳居住地发展提供了重要的参考价值。

Humans In its broadest sense the Vertical Forest was created as a "home for trees inhabited by people". The first Vertical Forest is also a complex experiment in cohabitation between trees and humans: more than 700 trees and about 480 humans live there as well as over 1,600 birds and insects.

>> see urban forest, sustainability

人类 广义地说，垂直森林是一个"有人类居住的树木之家"。首个垂直森林是树与人类共栖的复杂实验项目：有 700 多棵树和大约 480 位居民，以及 1 600 多只鸟和昆虫共同栖居于此。

>> 参见：城市森林，可持续性

Hundertwasser, Friedensreich Born Friedrich Stowasser (Vienna, 15th December 1928 - February 19th, 2000), he was an Austrian painter, sculptor, architect and ecologist. A mulit-faceted and unique personality, he anticipated different design elements and concepts currently found in the Vertical Forest through his works. Hundertwasser preached the idea of a new biological architecture, based on the presence of trees in houses (Die Baummieter – the Tree tenant), capable of establishing a fixed relationship between the number of people and the number of trees present in the lived space. In Milan, on the night between 27th and 28th September 1973 during the course of La Triennale, Via Manzoni in the city centre was closed to traffic in order to allow the lifting of 15 trees by a crane. It was a demonstration of tree allocation conducted by Hundertwasser: the trees were planted in several of the apartments along the road, protruding from the windows and clearly visible from the outside.

百水先生 原名佛登斯列·汉德瓦萨（生于维也纳，1928 年 12 月 15 日—2000 年 2 月 19 日），是一位奥地利画家、雕刻家、建筑师和生态学家。他

有着多重却独特的个性，关于垂直森林的许多设计元素和概念，都有被他预见到并在其作品中呈现。百水先生所倡导的新的生物建筑学中就提出建筑与植物共建的理念（树的租户）。他要在具有生命的空间里建立起人与树固定的数量比例关系。1973 年 9 月 27 日和 28 日晚上，也就是米兰三年展期间，市中心的曼佐尼大街被封锁了交通，为的就是方便吊车吊植 15 棵树木。这些树木被种植在沿路的几个公寓内，树木的枝干从窗户里长出来的景象从户外清晰可见。这是百水先生指导下的实践，示范了树木的新的生长空间。

Implantation (criteria of) Unlike what happens in traditional architecture, in the Vertical Forest the compositional, morphological and technical criteria defining the external appearance are intertwined with similar criteria regarding the installation design of the plant system. In the first Milan Vertical Forest the placement of plants was established by taking into account their specific needs in terms of exposure - evergreens on the south-west side, deciduous on the north-east -, development in height in relation to the design of the balconies and of the impact on the level of comfort in the apartments.
These criteria have informed and led the composition project itself, developed on the base of the ornamental and formal qualities of the plants, and taking into account the image provided by the facades both as a whole and in the areas corresponding to individual residential units. The choice of vegetation to be planted was made by a group of botanists and was carried out over two years. The selection was made based on the combination of two primary parameters: the aesthetic and physical properties of each plant - especially the height of growth of the stem and the width of the development of the foliage - and the relative positioning provided on the facade.

>> see mute architecture, nursery, plants-selection of

种植（的准则） 与传统建筑不同，建筑外形除了取决于功能构成、形态要求和技术规范这三个设计原则外，垂直森林的建筑外观还与植物系统的安放设置要求密不可分，因而对植栽也有一套相应的设计原则。在米兰首个垂直森林中，植物的布局考虑到了它们对于阳光的要求，因此常绿植物位于西南方布置，落叶植物则种在东北方。此外，植物的生长高度与阳台的设计和公寓使用舒适度被并置在一起进行综合考虑。

利用植物的观赏性及形式感，通过将建筑整体立面和居住单元局部的立面综合起来设计，该项目逐步得以完善。植物的配置是由一组植物学家共同决定的，用了 2 年多时间来实施。植物的选择基于两个主要参数，一个是每种植物的美学意义和物理性质，尤其是植物生长高度与冠幅，另一个是植物在建筑立面上的位置关系。

>> 参见：沉默建筑，苗圃，植物配置

Irrigation System The first Vertical Forest has a centralized "intelligent" drip irrigation system, designed according to the needs, distribution and placement of the plant organisms present inside it. The installation is equipped with a device for independent management for small groups of pots, which allows the adjustment of the water consumption depending on the real needs of the vegetation; needs which may vary significantly depending on the exposure and the height of each plant. A series of probes remotely controlled by a computerized device, carries out the monitoring of the humidity of plants: depending on to the data collected the water supply is activated or closed and the same sensors can detect malfunctions. The water system does not use drinking water but recycles that resulting from other uses (waste water or water from air conditioning systems with the use of ground water). Maintenance is centralized: a single unit is capable of managing the system and works according to a five-year plan.

>> see maintenance system, soil

灌溉系统 首个垂直森林使用了集中“智能”滴灌系统，根据植物有机体的需求、分布和定位设计。灌溉系统有一个可以独立管理一小组种植盆的设备，这样可以根据这组植物的实际需求来调整耗水量。每一株植物因为光照条件和生长高度的不同，其需水量也会有明显差异。一系列由电脑设备远程控制的探测器被用来监控植物的湿度。根据收集到的数据，供水系统可以灵活地开关。系统的故障也可以通过感应器探测出来。灌溉用水并非饮用水，而是将其他用水循环利用，如空调系统余水、地下水或废水。养护采用集中式管理，由一个独立的模块根据 5 年计划来运行这个系统。

>> 参见：养护系统，土壤

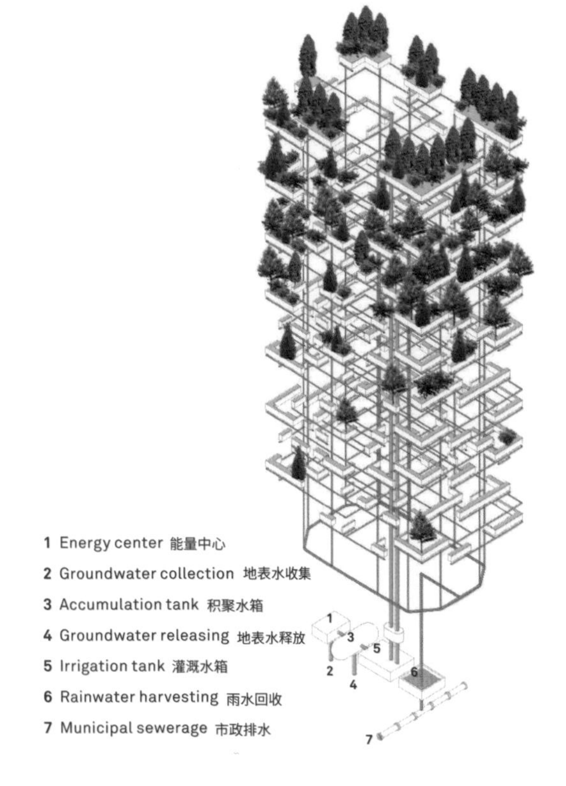

1 Energy center 能量中心
2 Groundwater collection 地表水收集
3 Accumulation tank 积聚水箱
4 Groundwater releasing 地表水释放
5 Irrigation tank 灌溉水箱
6 Rainwater harvesting 雨水回收
7 Municipal sewerage 市政排水

Italian Sparrow [*Passer Italiae*] A common bird of the Passeridae family, it is an established cross-breed between the Passer domesticus and Passer hispaniolensis, recognized as a species in 2013. It is found throughout Italy (except Sicily and Sardinia), southern Switzerland, Corsica and Crete. It can nest at up to over 2,000 metres above sea level, but is absent from valleys inhabited only in summer. It is omnivorous and closely dependent on man, thus it is usually found only in permanently inhabited towns and cultivated countryside. The nest is built in spaces under rooftiles, in holes in the walls, between bridge supports or more rarely in the hollows of trees.

意大利麻雀【意大利麻雀】一种普通的雀科鸟，是家麻雀和黑胸麻雀的杂交品种，这个品种于2013年被新近确认。它们在整个意大利（除了西西里岛和撒丁区）、瑞士南部、科西嘉岛和克里特岛较为常见。意大利麻雀筑巢于海拔2 000m以上，只有在夏季，才无法在山谷找到它们的踪迹。这种麻雀是杂食类鸟，与人类关系密切，因此可见它们出没在长期有人居住的城镇和耕种作物的乡村。这种麻雀喜欢在屋顶瓦片间、墙洞里和桥架下筑巢，很少会在树洞里筑巢。

Jackdaw [*Corvus Monedula*] A passerine bird belonging to the Corvidae family, found over a wide area stretching from northwest Africa across Europe, Iran, North-West India and Siberia. It lives mainly on steppes, in forests, fields, cliffs, but is also at home in populated areas, where it can prey on the eggs and young of pigeons. It nests within walls and rocks, large cavities with a relatively narrow entrance, and often coexists with other species. It has a typically fast and straight style of flight with deep wing beats.

寒鸦 【寒鸦】 属于鸦科的一种雀形目鸟。广泛分布在从非洲西北部到欧洲、伊朗、印度和西伯利亚西北部的地区。它们主要栖息在草原、森林、田野和悬崖，也会在有人口的地方栖息，因为有机会捕食到蛋和雏鸽。它们筑巢于墙和岩石中，以及具有较狭窄入口的大空洞中，通常可以与其他物种共存。它们飞行速度非常快，习惯于直线飞行，翅膀振幅大。

Jasminum mesnyi Hance [*Jasminum primulinum*] An evergreen shrub native to China, it can reach 5 metres in height and is found in gardens and public spaces, as a trailing plant or climber for small surfaces. It does not tolerate the cold and grows outdoors where the climate is mild and may be subject to partial defoliation in particularly cold winters. It is not found in the wild even in its area of origin, which is why there is speculation that it has been artificially cultivated for a considerable period of time.

野迎春 【云南黄馨】 原产于中国的一种常绿灌木，高度可达5m，常用于花园和公共空间，是一种枝条垂挂或可在小范围攀爬的植物。它耐寒性差，适宜生长在气候温和的户外。特别寒冷的冬季会导致部分落叶。在野外甚至原产地都难觅其踪，因此很长一段时间都有人断言说这种植物是人工培植的。

Kestrel [*Falco tinnunculus*] Among the most common raptors in central Europe, the kestrel often chooses the city as its natural habitat. It is characterized by its particular style of flight referred to as "Holy Spirit", during which it remains totally still in the air, with small wingbeats and with the tail held in the fan position, harnessing the wind to remain stable and looking at the ground in search of prey.

红隼 【红隼】 中欧地区十分常见的猛禽。红隼经常选择城市作为其自然栖息地。它以“圣灵”般的飞行风格为特色，能够在空中完全凝滞，一边小幅度地振动翅膀，一边把尾巴摆成扇状，利用风力来控制平衡，搜索地面、寻找猎物。

Kuhl's Pipistrelle [*Pipistrellus Kuhlii*] A bat of the Vespertilionidae family, widespread in North Africa, in Europe from Portugal as far as Kazakhstan, the Arabian Peninsula and from the Middle East to India. Small in size (the length of the head and the body varies between 35 and 55 mm), it takes refuge in often numerous colonies, many

in the interstices of buildings, bat boxes or more rarely in tree holes, rock crevices, caves or mines. In the northernmost of the areas in which it lives, it hibernates from November to March or April while in other parts it is active throughout the year.

古氏蝙蝠 【古氏伏翼】 蝙蝠科的一种蝙蝠，广泛地分布在北非、从葡萄牙到远至哈萨克斯坦的欧洲、阿拉伯半岛，以及从中东到印度的地区。它们体型非常小，头部和身子长35～55mm。它们将大部分聚居地都当成是避难所，聚居地常在建筑缝隙中、蝙蝠箱或罕见的树洞、岩石缝隙、洞穴或矿区。在其最北端的生活区，它们会在11月到次年的三四月间冬眠，而在其他地区，它们却全年活跃。

Laboratory-roof In the Vertical Forest the roof is designed as a large communal area to be used for its unique features in terms of studying the behaviour of animal species, implementation of biodiversity, the installation of devices for energy and environmental sustainability (wind turbines, photovoltaic panels). In the first Vertical Forest, parts of the floating flooring and modular tanks set above the ventilation ducts and installation roof fans have been designed to accommodate autochthonous herbaceous vegetation, with limited substrate thicknesses and with virtually zero maintenance needs. The washed gravel inside the tanks created along the ventilation ducts can be replaced with a high particle size substrate and with herbaceous flowering species (wild roofs). The presence of a green roof lowers the cover surface temperature: a fact that helps to optimize the yield of the solar panels, whose efficiency decreases at high temperatures.

>> see biodiversity, energy device

实验室屋顶 在垂直森林中，屋顶被设计为一个具有独特功能的大面积的公共区域，在此可以研究不同种类动物的行为，实践生物多样性策略，安置能源环境的可持续性设备（风力涡旋机和光伏板）。在首个垂直森林，通风管和屋顶通风扇的上方设计了部分挑空的地面铺装和水池，这样的设计可以在有限的基层厚度、零养护的条件下生长草本植物。沿着通风管布置的水箱内的砾石，可以使用较大粒子的基质以及开花草本（野生植物屋顶）来替代。绿色屋顶的存在降低了屋面的温度，实际上，因为太阳能板的效能在高温环境下会降低，绿色屋面还有利于优化太阳能板的效能产出。

>> 参见：生物多样性，能源设备

Ladybugs/Ladybirds The Vertical Forest is a biodiversity laboratory that lends itself to new forms of experimentation in terms of forms of interaction between man, nature and architecture. On May 29th 2014, the first Vertical Forest in Milan was the scene of one of the first experiments in the field. On that occasion, 1,250 ladybirds of the species adalia bipunctata were released on the terraces and the roofs of the two towers. The action of inserting these beetles had a dual role: on the one hand it allowed the scientific group to obtain a wide range of new data about the behaviour of these animals at a height of over 100 metres, and it also ensured effective ecological protection

for the plants against any invasion by aphids (parasitic species that feed on sap, thus damaging the vegetation).

>> see biodiversity, maintenance system

瓢虫 / 瓢甲科 垂直森林是一座生物多样性的实验室，它是一种新型的实验方式，可以研究人类、自然与建筑之间的互动形式。2014 年 5 月 29 日，米兰的首个垂直森林成为了这一领域的新实验之一。那天，1 250 只红色黑点的瓢虫被释放到两栋大楼的平台上和屋顶上。这个行为有双重作用：一方面，科学团队可以获得关于在 100m 高处这些动物行为的大量新数据，另一方面，它们可以为植物提供有效的生态保护，防止植物受到任何蚜虫的侵害（通过吸取树液生存的寄生物种，会导致植被被破坏）。

>> 参见：生物多样性，养护系统

Leadwort [*PCeratostigma plumbaginoides*] A species of the Plumbaginaceae plant family, native to the temperate and tropical regions of Asia

and Africa. With a height from 0.3 to 1 metre, its leaves are simple, up to 9 centimetres long and arranged spirally. The flowers have a five-lobed corolla ranging in colour from dark-blue to red-purple; the fruit has a small capsule that carries only one seed.

山灰柴 【蓝雪花】 蓝雪花科的一种，原产于亚洲和非洲的温带和热带地区。高度在0.3～1m，它的叶子构造简单，长度可达9cm，呈螺旋状排列。花有五个分裂的花冠，颜色从深蓝色到红紫色，有一个只装有一颗种子的小蒴果。

Little Ringed Plover [*Charadrius Dubius*] A migratory bird of the family Charadriidae, it is found all over the world except in the Americas. It lives in gravelly or stony areas near rivers and freshwater streams. It nests on rocky or stony ground with little or no plant presence.

黑领鸻【金眶鸻】 鸻科的一种候鸟。除了美洲，在世界上其他地方都非常常见。它们生活在靠近河流和淡水溪流附近的碎石或多石区，在植物较少的碎石或多石地筑巢。

Maintenance system Similar to a 'live' system that is evolving over time, the Vertical Forest highlights the need for targeted programmes of maintenance and control of the greenery, planned right down to the smallest details. The vegetation of the first Milan Vertical Forest is regularly taken care of through periodic actions of various kinds (pruning, fertilizing, plant protection monitoring and anchor systems), managed directly from the Porta Nuova property and entrusted to specialized companies. The maintenance programme adopted consistently delivers economies of scale compared to piecemeal management of the task. Checking the health of trees, shrubs, ground conditions and the root systems is provided by teams of pruners, specialized maintenance groups and agronomists; the make-up of the experts and technicians of the teams varies depending on the season and type of action required. All the operations are carried out either from inside or outside balconies; the outside operations are conducted thanks to purpose-built cradles supported by two fixed cantilever crane supports positioned on the roofs of the towers. The maintenance programme includes the carrying out of two annual external operations for pruning and two/four annual internal operations by means of access to the apartments by a pair of specialized workers. Once a month, between May and October, a monitoring and check of the phytosanitary progress and gen-

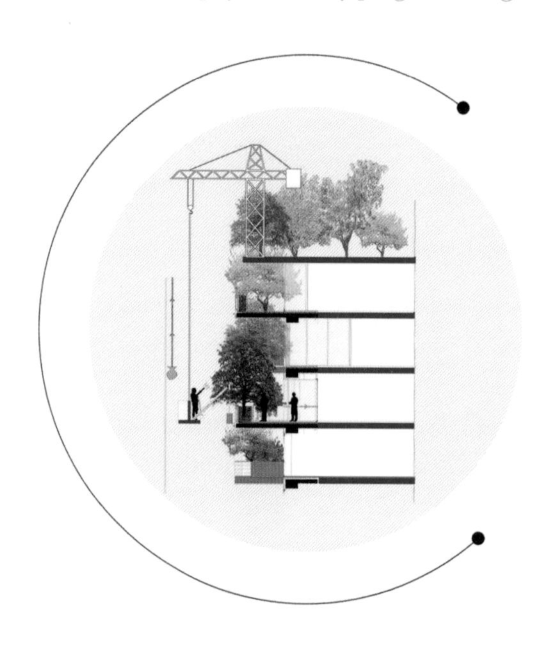

eral status of the plants is also carried out. Any pest attacks are controlled biologically.

>> see ladybugs/ladybirds, irrigation system

养护系统 类似一个随着时间进化的“活”系统，垂直森林非常强调对绿色植物的养护计划，对绿色植物的控制深入到最小的细节之中。米兰首个垂直森林中的植栽，都得到各种定期的照料（如修剪、施肥、植物保护监控与锚固系统等），由新门地产直接统筹委托给专业公司管理。相比于细碎的管理，整体养护体现出了规模经济效益。检查乔灌木健康情况、场地条件、植物根系抽检等工作都由专业的修枝机械团队、专业的植物养护团队和农学家负责。随着季节变化和需要，这个团队的专家和技术人员构成也会发生变化。所有的操作都是在阳台内外同步进行的。阳台外的操作通过具有特殊用途的吊篮辅助进行，吊篮由两个放置在大楼顶部的固定悬臂吊车支撑。阳台内的维护项目包括两位专业人员进入公寓内，提供每年两次的植物外形修剪，及每年两次或四次的养护操作。从 5 月到 10 月，每个月进行一次植物检疫，并检查植物状态。所有灭害虫的行为，都是通过生物方法进行的。

>> 参见：瓢虫 / 瓢甲科，灌溉系统

Manna Ash [*Fraxinus ornus*] Tree or shrub of the Oleaceae family, it is widespread throughout southern Europe and Asia Minor. The northern limit of the species is indicated by the Alps and the Danube valley; the eastern by Syria and Anatolia. Very common in Italy throughout the whole peninsula from 1,000 metres above sea level in the north to 1,500 metres above sea level in the south, it is almost absent in the Po Valley. Resistant to harsh weather conditions, it is suitable for the reforestation of deserts and drought-stricken areas and is grown for the production of manna, or in vineyards as a support for rows of vines. Reaching a height of up to 4-8 metres, it reproduces easily with planting.

欧洲白蜡木【苦枥木亚属】木犀科树木或灌木，广泛分布于欧洲南部和小亚细亚地区。这些物种北部的界限为阿尔卑斯山脉和多瑙河河谷；东部为叙利亚和安纳多利亚。在意大利从北部海拔 1 000m 以上地域到南部海拔 1 500m 以上地域的整个半岛都很常见，但在波河流域几乎没有。花白蜡树对于恶劣的天气具有抵抗力，它适合作为荒原和干旱灾区的重新造林，且已用于吗哪的生产，或可以在葡萄园中作为葡萄的支架。这种树可以高达 4～8m，非常容易栽培。

Materials The interaction in the Vertical Forest between mineral and vegetable architectural components opens unprecedented possibilities for experimentation on materials and finishes. In the first Vertical Forest created in Milan, the theme is explored through numerous design variants, including facade solutions based on wood paneling and exposed concrete. It was possible to observe a genuine archetypal draft version during construction of the latter version of the project, which had a strong plastic impact when the buildings had not yet been equipped with the ceramic cladding panels.

>> see colour, three dimensional facade

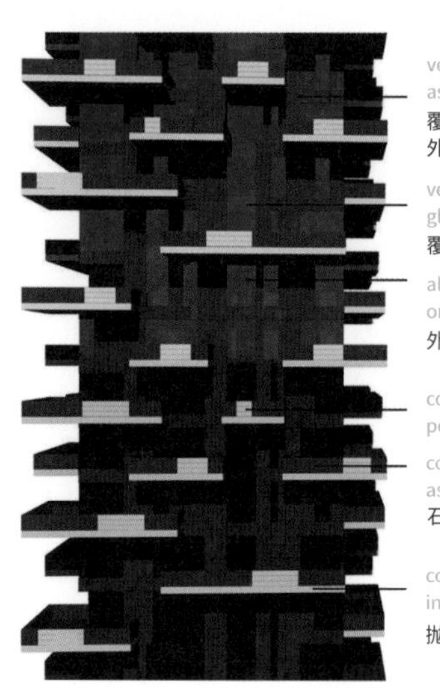

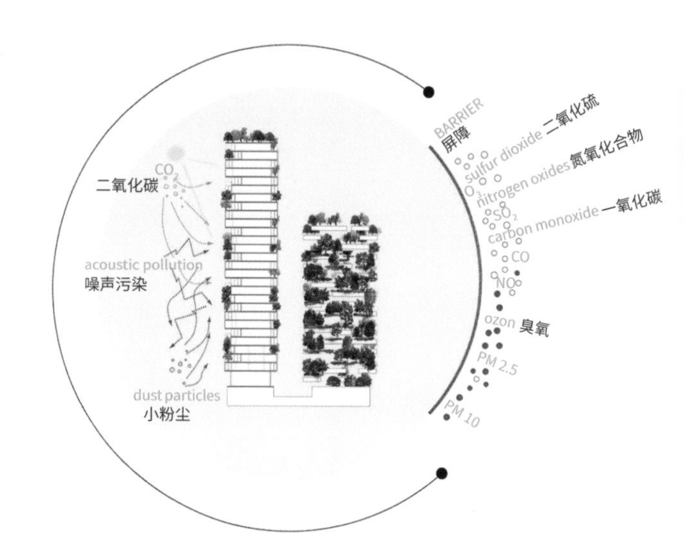

材料 垂直森林中矿物表皮和植物材料相互作用，这为建筑材料的试验开辟了前所未有的可能。在米兰的首个垂直森林，通过大量设计变化对这个主题进行了探讨，其中包括了木墙板和清水混凝土配合的立面组合。在项目的后期施工期间，还未安装陶瓷面砖时，可以看到原先草图的设计（清水混凝土）所具有的强大的形体冲击力。

>> 参见：颜色，三维立面

微气候 在垂直森林中，因每个公寓大小不同，植物作用下的微气候和场景也有差异。这些植物过滤器有助于减少建筑物的能源消耗（大约 30%），可以吸收二氧化碳（就垂直森林这个案例来说，大概每年吸收约 19.825 kg 的二氧化碳），可以过滤掉空气中细粉尘粒子，减轻噪声污染，避免阳光直射和直接风吹，并且还可以产生氧气和水分。

>> 参见：三维立面，可持续性

Microclimates The presence of plants in the Vertical Forest determines specific micro-meteorological and micro-climatic scenarios which vary depending on the dimensions of each apartment. The plant filters help to reduce the energy consumption of buildings (about 30% less), they absorb CO_2 (estimated to be about 19.825 kg / year in the case of the first Milan Vertical Forest,), they filter out fine dust particles present in the air, mitigate noise pollution, protect from direct sunlight and wind and produce oxygen and moisture.

>> see three-domensional facade, sustainability

Mineral city The idea of creating a tower completely surrounded by trees was conceived in 2007 in Dubai by Stefano Boeri, then director of Domus, who thus described the frantic construction of one of the cradles of the new oil and financial capitalism, "a mineral city, made up of dozens of new towers and skyscrapers, all clad in glass or ceramic or metal, all reflecting the sunlight and therefore acting as heat generators in the air and especially on the ground inhabited by pedestrians". Another influence on the Vertical Forest's nascent project was an article by Spanish architect Alejandro Zaera-Polo, published in the same period in the journal of the School of Design at Harvard University (High-Rise Phylum, on: Harvard Design Magazine No. 26, Spring / Summer 2007). This article provided a reading of the un-ecologically sustainable effects of curtain wall architecture, revealing in addition that about 94% of the tall buildings in the world built after 2000 were covered in glass. So in its initial conception, the Vertical Forest was therefore created in opposition to the "mineral city" of steel and glass, well represented in Dubai and effectively described by Zaera-Polo, with a new model of high level building able to incorporate plant life and to merge the principles of sustainability and biodiversity.

>> see biodiversity, demineralization, energy device

矿物城市 建设一栋完全由树木包裹的大楼，是斯坦法诺·博埃里于2007年在迪拜产生的一个设计构想。当时他是Domus杂志的主编。他将迪拜这个新石油和金融资本摇篮之一的奇妙的建设描述为："由许多新的塔楼和摩天大楼组成的矿物城市。所有的建筑都覆盖着玻璃、水泥或金属，所有这些都会反射太阳光，并作为发热器，尤其会提高地面和街道空气温度。"

垂直森林的初期，还受西班牙建筑师阿莱桑德罗·扎埃拉·波罗同期发表在哈佛大学设计学院期刊上的一篇文章的影响（《高层之门》，发表于哈佛设计杂志第26期，2007年春季／夏季）。这篇文章对幕墙架构的非生态可持续性影响进行了分析，其中提到，在2000年后所建设的高楼中，约94%的高楼立面被玻璃覆盖。所以，在概念初期，相对于覆盖以钢铁和玻璃的"矿物城市"的建筑，正如迪拜的建筑所呈现的，垂直森林是完全相反的概念。扎埃拉·波罗将植物生命、可持续性及生物多样性原则融合在高层建筑的新模型中，对概念进行了有力的描述。

>> 参见：生物多样性，去矿化，能源设备

Mute architecture The Vertical Forest takes a picture of a state of uncertainty in traditional architectural practices, and looks at the need to revitalize the language of design through the use of external elements in which nature takes its place as size, shape and material. In opposition to the idea of 'shouted' architecture, the architecture of the grand gesture which circumscribes, tames or puts "green" on a pedestal, the Vertical Forest speaks the elementary architectural language of 'silent' architecture, which takes second place in order to support the growth and the imperfection of plant life.

>> see biodiversity, basic radicality, villas

沉默建筑 垂直森林是传统建筑建设中不确定性的一个呈现。它通过使用一些外在元素，来满足激活设计语言的需要，比如，以自然物表达大小、形状和材料。有些"叫嚣"的建筑，唯我独尊，将"绿色"限定、抑制，或踩在脚下。垂直森林恰恰相反，它讲述了一个"无声"建筑的基本语言，它生来就居于次席，支撑起植物的成长与生命的瑕疵。

>> 参见：生物多样性，激进的原型，别墅

Nightingale [*Luscinia megarhynchos*] A migratory passerine bird of the Muscicapidi family, measuring about 16 cms. It is widespread and very common in Asia, Europe and North Africa and it can be found in dense deciduous forests or thickets and particularly prefers moist soil. It nests close to the ground and its singing, considered among the most beautiful among the songbirds, consists of single and dual tones densely aligned with one another. In the cities it is often forced to sing at a greater volume to overcome the noise. In early spring the singing takes place mostly at night and serves to indicate territory and attract females.

夜歌鸲【夜莺】 鸲科的一种雀形目鸟类，身长约 16cm。它在亚洲、欧洲和北非很常见，在茂密的落叶林或灌丛中可以看到它们。夜莺尤其喜欢湿润的土壤，筑巢一般接近地面。夜莺的歌喉是由单音调和双音调搭配起来的，被认为是世界上鸣禽中最美妙的声音。在城市里，为了克服噪声问题，它们不得不大声歌唱。在早春，夜莺一般都是在晚上歌唱以宣告它们的领土权，并吸引雌性夜莺。

Nursery The presence of the plant components in the Vertical Forest makes it more similar to a set of processes rather than an architectural object tout court; this set of processes can be articulated over time and which may precede or follow the birth of the building itself. One of the most important of these is the process regarding the cultivation and selection of trees to be planted, which in order to best adapt to the growth conditions found in the Vertical Forest must develop specific qualities in terms of size, structural strength and a guarantee of overcoming the post-transplant phase. In order to achieve this, for the first time in Italy, in the case of the Milan Porta Nuova Vertical Forest a two year pre-cultivation contract was signed. During the summer of 2010, the plants intended to be planted later on the two towers were taken from their nurseries of origin and placed inside special recyclable plastic air-pot containers located in the Peverelli nursery near Como: the botanical 'nursery' chosen by the contracting company. The material, dimensions and mechanical characteristics of the air-pots were specifically calibrated to allow them to adapt to the growth of the trees, and at the same time to promote the formation of a root system in the plants with the required size, characteristics, balance and ample supply of radicles . During the growth period in the 'nursery', the Vertical Forest project group experts were able to verify and in some cases correct the structural conformation of the trees in order to obtain a foliage format that would be most suitable for the final growth environment.

>> see plants - selection of

苗圃 垂直森林里植物的存在，使得它看起来更像是一套流程，而不是一个简单的建筑个体。其中最重要的过程之一，就是选择需要培育和种植的树。为了最好地与垂直森林中所能提供的生长条件相适应，必须提前就植物的大小、结构强度，以及为克服移植后的问题等订制方案。为了实现这个方案目标，在意大利首次签订了关于米兰垂直森林的两年预栽培合同。在2010年夏天，准备种在两栋大楼中的植物从原产地的苗圃中被取出，然后种在了距离科摩较近的佩维莱立苗圃内，放在专门的可回收塑料空气盆当中。植物"苗"是由签约公司选择的。空气盆的材料、大小和机械性能都特别经过校准，从而使得它们可以与树木的成长相适应，这样也促进了植物根系的形成，使根系的尺寸、性质、平衡和根系供给达到项目要求。植物在"苗圃"的生长过程中，垂直森林的项目组专家能够对其形态结构进行验收，在某些情况下，他们还纠正植物的形态结构，使其最适合于其最终生长环境。

>> 参见：植物配置

Pallid Swift [*Apus pallidus*] A bird of the Apodidae family, lighter in colour than the common swift but with a similar song. It has a large natural habitat, estimated at between 1,000,000 and 10,000,000 km², even if it reproduces only in southern Europe. Throughout the entire European continent, the population is calculated at between 77,000 and 320,000 specimens: an amount that testifies to its classification as a low-risk species. It chooses to nest mainly in slits or holes in the exterior walls of buildings.

苍雨燕【苍雨燕】 雨燕科的一种鸟，比一般雨燕的毛色要浅，但是，歌喉相似。它的自然栖息地范围较大，估计在1 000 000～10 000 000km^{2}之间；

但它仅在欧洲南部繁殖。据测算，在整个欧洲大陆苍雨燕的数量在77 000到320 000之间，这个数据可以表明苍雨燕是一种低危物种。它喜欢在建筑外墙的缝隙或者洞里筑巢。

Parrot [*Psittaciformes*] The common name of parrot is used to refer to Psittaciforms, an order of birds including several neornithe species. Divided into three families, the order is very widespread in tropical and subtropical areas throughout most of the planet: the greatest level of biodiversity is found in South America and Australasia. The bright colours which almost all parrots have, along with the ability to emit 'spoken' sounds, make them very popular as pets.

鹦哥【鹦鹉】 鹦鹉通常指的是吸蜜鹦鹉，按照鸟类排序，包括好几个鸟亚纲物种。鹦鹉被分为三个家族，在地球上，大部分热带和亚热带地区广泛

分布，生物多样性水平最高的地区当属南美和澳大利亚。几乎所有鹦鹉的羽毛颜色都非常鲜艳，它们可以发出“说话”声，所以它们是非常受欢迎的宠物。

Plants (selection of) The unique growing conditions in the Vertical Forest require a choice of trees with precise characteristics. All the medium and large plants planted in the first Vertical Forest Milan have been selected according to a combination of aesthetic (related to their ornamental potential) and technical (related to safety and maintenance over time) criteria. In particular this evaluation took account of the structural strength of the trees and of their innate ability to adapt, or in other words, the ability to tolerate external stress and pruning in order to maintain foliage without losing their natural appearance. Among other characteristics of the plants of the Vertical Forest is the presence of a non-aggressive root system, reduced allergenic potential, efficiency in micro-anchoring, ease of maintenance and resistance to disease.

>> see Wind tunnel, Implantation (criteria of), Nursery, Soil

植物配置 垂直森林独特的生长条件，使其对植物的选择有某些特别要求。米兰垂直森林里种植的所有大中型植物都综合了审美标准（观赏价值）和技术（与安全和维护相关）标准。在评估当中，考虑到了树木的结构强度以及它们天生的适应能力，换句话说，即它们对外部压力和修剪的耐受能力。在不失去其天然外观的条件下，可以对它们进行修剪。在所有垂直森林的植物特性当中，非侵略性根系可以减少潜在的致敏性，还可以有效地进行微锚固，并且具有抗病能力，较容易养护。

>> 参见：风洞，种植，苗圃，土壤

Plant species The first Vertical Forest serves as a habitat for more than 100 different plant species, more than those commonly found in a public park in the neighbourhood of about 5,000 m^2.

植物品种 垂直森林是 100 多种植物的栖息地。其中大部分品种在约 5 000m^2 范围内的公园内都很常见。

Pots The plant pots themselves are a project-in-a project of the Vertical Forest. In the first version of the Milan Vertical Forest, these items are placed along the peripheral edges of all balconies, with a linear development of approximately 2 kms, different depth levels (30 to 110 cm) and a constant height (1 metre 10 cm). The size of the pots was calibrated to ensure the appropriate conditions for the development of the plants while considering the specific parameters involved in their nourishment and the satisfying of their water requirements. Made of concrete, the pots have been designed to ensure water resistance and drainage ability, as well as preserving the specific features of the soil, which should not be too 'loose' in order not to adversely affect the stability of the turf. The substrate in which the roots grow is maintained at a distance from the waterproofing membrane on the bottom of the pots through a separation and drainage layer, consisting of two filters of non-synthetic fabric: a root barrier in low-density PE (very ductile) and a particular geo-textile which isolates the roots from the walls through the formation of a sort of air chamber.Within the pots there are fixing systems for attaching the metal anchorage bars of the plants, designed not to adversely affect the waterproofing abilities of the pots themselves. On the bottom there is a separation layer and a steel welded mesh, which acts as an anchor for the earth in which the roots develop. For larger trees an additional anchoring plane was prepared, based on steel cables, able to adapt over time to the flexibility and growth of the plant. Management of the pots is subject to building regulations, as with maintenance of the greenery and the number of plants per pot.

>> see anchor systems, balconies, irrigation, maintenance, structure, soil

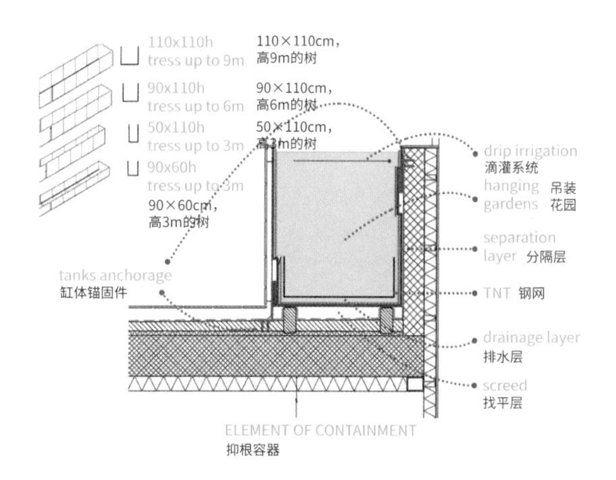

种植盆 种植盆本身就是垂直森林这一项目中的项目。在米兰垂直森林的第一个设计版本当中，它们呈线形沿着阳台外围摆放，长度可达约2km，深度不一（30～110cm)，且高度恒定（1.1m)。在考虑到植物营养所需的参数并满足浇灌要求的基础上，种植盆的大小是标准化的，可以为植物种植提供适宜的条件。种植盆是用混凝土做的，以保证其防水性能和排水性能。它能保留土壤的某些特性，比如土不能太“松”，避免对草皮的平整稳定带来不利影响。植物根系生长的末端与树盆底部的防水膜有相当一段距离，中间有个隔层和排水层。排水层是由2个非合成纤维过滤器组成：一个低密度PE（延展性特别好）的根障和一个通过形成气囊的方式将植物根部与墙面隔离的土工织物。在种植盆里有一些将植物固定在金属紧固杆上的固定装置，这种设置不会对种植盆本身的防水能力产生负面影响。在底部，有一个隔离层和一个钢制焊接网，用来锚固根部土球。为了固定较大的树木，还额外增加了一个用钢索组成的锚定平面，可以使植物更好地适应和生长。种植盆的管理必须根据建设规程来进行，比如说每个种植盆中绿色植物的养护以及植物数量的控制。

>> 参见：锚固系统，阳台，灌溉，养护，结构，土壤

Redstard [*Phoenicurus Phoenicurus*] A small migratory passerine bird, attributed to the Muscicapidae family. It winters in the tropical countries from the Red Sea to the African lakes and during the warm seasons lives throughout all of Europe, but also in North Africa and more rarely in the islands. The male reaches the warm places first in early April, often a few days ahead of the female. It lives in woods and in public parks, especially where there are trees with many cavities that are used by the females to build nests with a typical amphora shape.

欧亚红尾鸲 【欧亚红尾鸲】 一种迁徙类雀形目鸟，属于鹟科。它从红海来到非洲湖泊地区的热带国家越冬，而在温暖的季节，则选择在欧洲和非洲北部，以及少数岛屿聚居。雄性红尾鸲会先于雌性鸟几天在 4 月初到达温暖的地方。红尾鸲居住在树林和公园，尤其是有很多空洞、可供雌鸟筑巢的树上。

Replicability The Vertical Forest has also been designed to be a variable and replicable model, able to form the basis for a series of interconnected interventions. This attribute is due directly to its ability to act as an attracting element for biodiversity and as a driving force for new urban centres. One example is the Vertical Forest created at Porta Nuova in Milan, close to the important infrastructure of the Garibaldi Repubblica area and designed as the first of a potential sequence of grafts in areas located near major underground stations in the city. The Vertical Forest model can also be adapted to different types of intervention public buildings or private residential structures, tertiary sector, representational or public structures- and can add value to abandoned buildings or those whose construction has been interrupted.

>> see biodiversity, urban traffic

可复制性 垂直森林被设计成一个可变化且可复制的模型，它可以为一些相互关联的行为提供基础。这种属性，直接归因于其在吸引生物多样性元素方

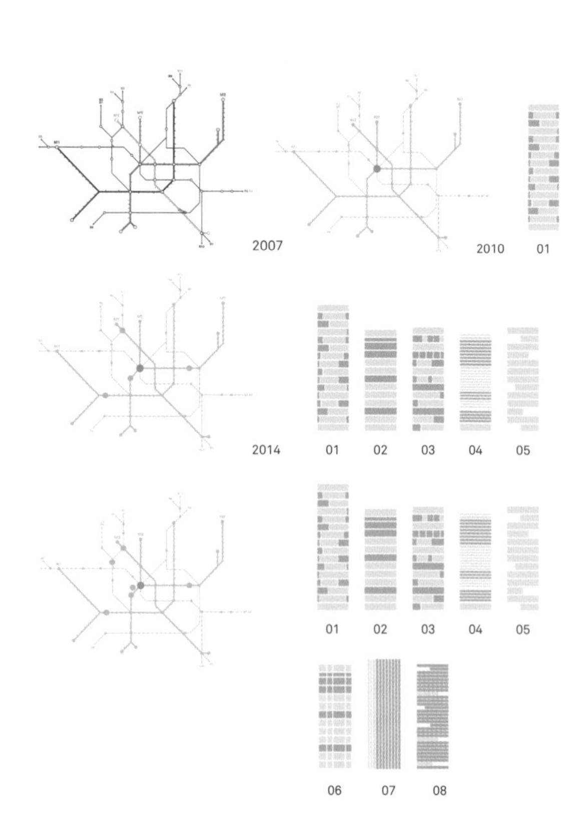

面的能力，以及它具备成为新城市中心的驱动力。其中的一个事例，是接近加里波第共和广场这个重要基础设施的米兰垂直森林，它被设计成为首个位于城市地铁站附近的植物移植模型。垂直森林的模型，适于作为公共建筑、私人住房、产业、代表性建筑或公共建筑物等不同类型。此外，它还可以为那些废弃建筑或中断施工的建筑增加价值。

>> 参见：生物多样性，城市交通

Robin [*Phoenicurus Phoenicurus*] A small passerine songbird of the Muscicapidae family, it is the only known species of the genus Erithacus. It is widespread throughout Eurasia and North Africa, in an area extending from the islands of the Azores to the west to western Siberia in the east. Its favoured habitat is coniferous forests, but it is often found in other types of woods and gardens. The nest, in the shape of a perfectly round cup, is made in holes or cracks in trees, at the foot of hedges or even in old objects abandoned by man. It has a distinctive song, consisting of a series of varied and well-defined short warbling phrases.

罗宾鸲 【欧亚红尾鸲】 属于鹟科的一种雀形目科鸣禽，它是已知的唯一一种欧亚鸲属。它广泛分布于欧亚大陆和北非，还拓展到从亚速尔群岛到西西伯利亚东部的地区。它偏爱的栖息地是针叶林，但也经常会在其他类型的树林和花园中出没。它经常在树洞、裂缝、树篱脚下，或被人们所废弃的旧物里筑巢，鸟巢呈圆形杯状。它的歌声与众不同，是由一系列不同的短颤音乐句组成。

Rose of Sharon [*Rose of Sharon*] A semi-evergreen shrub of the genus Hypericum, widespread throughout the Middle East and Eastern Mediterranean, especially in the Strandja mountains along the Black Sea coast. Often grown for its large yellow flowers with five petals, it remains rather small in size with a maximum height of 1 metre and 2 inches in width. The elongated leaves grow in opposing pairs.

沭阳木槿 【沭阳木槿】 一种半常绿灌木，广泛分布于中东和地中海东部，特别是沿黑海海岸的斯特兰贾山。大多数时候它的花为五瓣花瓣的大黄花，但其本身体量却挺小，其最大高度和宽度分别只有 1m 和 5.08cm。其对生的叶片呈狭长形。

Serin [*Serinus*] A bird of the Fringillidae (finch) family and a close relative of the wild canary. It is widespread throughout Europe, Asia, North Africa and prefers wooded countryside, parks and gardens. Very common in the temperate and warm regions of Europe, it does not venture far into the northern areas of the continent. In Italy it is more frequent in the Centre and South in the cold season; in the north in the summer and also in the Alps. Highly sociable, it forms small flocks of about a dozen specimens and often looks for food on the ground, where it jumps from place to place. It can be bred as an ornamental bird and to be re-introduced.

金丝雀【欧洲丝雀】 雀科的一种鸟，还是野生金丝雀的近亲。它广泛分布于欧洲、亚洲和北非地区，喜欢栖息于树木繁茂的乡村公园和花园。在欧洲的温带和温暖的地区很常见，它不喜欢欧洲大陆北部地区。在意大利的冬季，它更频繁地在意大利中心和南部地区出没；夏季，在阿尔卑斯山也很常见。这是一种好社交的鸟类，它们经常几十只一群，在地面寻找食物，并在地上跳来跳去。它是一种可以作为观赏鸟来养殖并引种的鸟。

Snowy Mespilus [*Amelanchier lamarckii*] A deciduous shrub belonging to the Rosaceae family, native to North America and widely naturalized in Europe. Very durable, it is often grown as an ornamental plant. It produces fragrant white star-shaped flowers and dark red sweet edible fruit which becomes dark purple when ripe.

拉马克唐棣【拉马克唐棣】 一种蔷薇科落叶灌木，原产于北美，广泛分布于北美和欧洲。非常耐用的树木，它通常被当成观赏植物种植。其花芳香，呈白色星状，结深红色甜果实，这种果实在成熟时会变成深紫色。

Soil The Creation of the Vertical Forest has offered the opportunity to develop a wide range of technological and design innovations, among which are those concerning the characteristics of the soil used for the growth of the plants. The substantial weight of the pots, combined with the notable overhang of the balconies, made it necessary to identify specific solutions that minimize the load imposed on the building - and therefore the size of the loadbearing structures. For this purpose a particular soil was identified that was very light but at the same time able to meet the nutritional needs of the plants. It is formed by a mixture of organic and inorganic materials, well drained and enriched with fertilizers. The dimension criteria combine the need to limit the volume of the pots with a guarantee of the best growth conditions for the trees, and can be summarized in a simple datum: one metre of thickness of the cultivation substrate.

>> see balconies, irrigation, plant selection, structure, pots

土壤 垂直森林土壤的研制为广泛的技术和设计革新提供了机遇，其中特别关注什么样的土壤才能满足植物生长所需。由于种植盆与阳台的悬挑部分的重量，设计人员有必要寻找一种可以最小化负载的建筑方案，因此，就必须考虑承载负荷的结构尺寸。为此，相关人员配制出了一种重量轻，却可以满足植物营养需求的特定土壤。它是由有机和无机材料组成的混合物，排水性强且肥料丰富。种植盆的标准尺寸，已经考虑到种植盆的容积受荷载的限制，同时确保能为植物提供最佳的生长条件。这个标准可以归纳为一个简洁的数据：能容纳 1m 厚的栽培基质。

>> 参见：阳台、浇灌系统、植物配置、结构、种植盆

Spotted Flycatcher [*Muscicacapa striata*] A small migratory passerine bird of the Muscicapidi family, measuring an average of 14 centimetres in length and 16 grams in weight. It nests in most parts of Europe, North Africa and western Asia. It usually winters in sub-Saharan Africa and south-west Asia. In Italy it is both resident and transitory, although in recent years there has been seen a significant reduction in its presence. It prefers sunny forests, parks, gardens and more generally open surfaces with numerous scattered trees.

斑鹟【斑鹟】 鹟科迁徙小雀形目鸟，平均长度为 14cm，重量平均为 16g。栖息于欧洲、北非和亚洲西部的大部分地区。通常会迁徙到撒哈拉以南的非洲和亚洲西南部过冬。在意大利，它既是常驻于此的鸟，也属于过客，但近几年来，斑鹟的数量明显减少。这种鸟喜欢阳光充足的森林、公园、花园，以及树木分散种植的绿色开放区域。

Star jasmine [*Rhynchospermum jasminoides*] A climbing shrub of the Apocinacee family, a native of Southeast Asia (Japan, Korea, China and Vietnam) but spread throughout different areas of the globe, particularly in the south-eastern United States. It is often grown as an ornamental and household plant, as well as in gardens and public spaces and can also be trained over pergolas. The leaves are evergreen, opposed and the flowers have five stamens placed on a white rotated corolla.

石崚【络石藤】 夹竹桃科的攀援灌木。原产于亚洲南部（日本、韩国、中国和越南），但是现在已分布在全球不同的地区，尤其是美国东南部。它可以作为一种观赏类植物种植在家里、花园或者公共场所，同时还可以缠绕在绿廊。其叶子是常绿的，白色花冠上有 5 个雄蕊。

Strawberry tree [*Arbutus unedo*] A shrub or small tree belonging to the Ericaceae family, widespread in the countries of the western Mediterranean and southern coasts of Ireland. It resembles a bushy evergreen shrub with new young branches being of a reddish color, and its height varies from 1 to 8 metres. The fruits ripen in the year following the flowering from which they grow: in autumn the presence of flowers and ripe fruit makes the plant particularly ornamental. It prefers acidic soils, and grows at altitudes between 0 and 800 metres above sea level.

草莓树【草莓树】 杜鹃花科的一种灌木或小乔木，广泛分布于地中海西部和爱尔兰南海岸。它与带有红色嫩枝桠的常绿灌木相似，高度一般为1～8m。它在开花期之后结果，秋天的花和成熟的果实让这个植物极具观赏性。它喜欢酸性土壤，生长于海平面到海拔 800m 的区间内。

Structures The structures or frameworks of the two towers of the first Vertical Forest are in reinforced concrete cast on site; the same material is used for floors, made of post-tensioned concrete. The post-stress technique allowed the slab thickness to be limited at equal flow rates, and thus a limiting in the amount of cement used in the construction. The foundations do not have any structural differences to the standard construction types. Developed by Ove Arup & Partners, the structural design provides a maximum internal span of about 10 m, with terrace overhangs of about 3.25 m (6 m for the corner terraces). The presence of pots on the edges of the balconies created considerable structural loads for the design: 11kN / m for those containing plants with heights of up to 3m and 13kN / m for pots with trees up to 6 m (+ 7kN / m for trees with a minimum base of 3m). The project also takes into account the increase in weight of a tree over time: it is estimated that the largest tree (6 metres high), without turf or soil, can during its life almost double its weight, so the project calculation is as follows: 300 kg of weight for the largest tree at time of planting and 600 kg (max) for the same tree during its lifetime. To these loads should be added the dynamic stresses induced by the wind. During the design phase, a geometric-dimensional study of the plant species selected was carried out in order to determine: the height of the trunk, the area and centre of gravity of the foliage and air permeability. The structural design also benefited from tests carried out in the wind tunnel on scale models and full size ones.

>> see wind tunnel, anchor systems, soil, pots

结构 首个垂直森林的两栋大楼的结构为钢筋混凝土框架，楼板地面为后张混凝土。后张技术用于限制同等重量下的平板厚度，由此限制了施工过程中水泥的使用量。地基，与标准的建筑相比，并不存在任何结构差异。奥雅纳公司的结构设计使最大内部跨度可以达到 10m 空间，阳台出挑约 3.25m（拐角处的露台可以出挑 6m）。阳台边缘的种植盆对这个设计形成了较大的结构负荷。如果种植盆种植植物高度为 3m，结构负载则为 11kN/m；如果种植植物高 6m，结构负载为 13kN/m。这个项目还考虑到了树木重量。随着生长，预计最大的树（6m 高）在不计算草坪或土壤重量的条件下，它的重量可以翻倍：种植时，最大树木的重量为 300kg，那么其存活期最大的重量为 600kg。风力所造成的动态压力也应当加到负载重量当中。在设计阶段，对所选植物还进行了几何－空间研究，据此确定出树干的设计高度、植物的重心位置以及空气渗透性。风洞中基于比例模型和实际模型基础上进行的测验也为结构设计提供了帮助。

>> 参见：风洞，锚固系统，土壤，种植池

Sustainability The Milan Porta Nuova Vertical Forest is part of a complex comprising 25 pre-certified LEED (Leadership in Energy and Environmental Design) buildings. This is the most important international instrument for energy and environmental certification of the design and implementation process. Even though the energy sustainability criterion is part of its founding principles, the Vertical Forest is nevertheless the result of a vision that puts the concept of biodiversity before that of sustainability. While the primary goal of sustainable architecture is to minimize its impact on the environment while always keeping a strong anthropocentric vision of the project, the approach to the concept of biodiversity within the Vertical Forest is based on the idea that mankind is just one of the many presences on the planet, and that is why new forms of cohabitation have to be found.

>> see biodiversity, continuous city, microclimates, basic radicality, humans

可持续性 米兰垂直森林是一个由 25 个具有 LEED（能源和环境先锋设计）预认证建筑组成的综合设施的一部分。LEED 认证是国际上关于能源和环保设计与实施最重要的一项认证。能源可持续性标准只是垂直森林基本原则的一部分，正是因为把生物多样性理念的想象放在比可持续性概念更重要的位置，才得以呈现现在的垂直森林。传统可持续性建筑的目标在于把建筑对环境的影响最小化，但仍然维护以人类为中心的观点。但是，在垂直森林，维护生物多样性的概念，是基于人类只是这个星球上众多生物的一种，这也是为什么人们需要找到可以与自然共栖的新形式的缘由。

>> 参见：生物多样性，连绵城市，小气候，激进的原型，人类

Swallow [*Hirundo rustica*] A small migratory bird of the Passeriformes order (Hirundinidae family), it is characterized by a long forked tail, sharp curved wings and a small straight dark grey beak. A widespread bird, (the global population of swallows is estimated to be at least 200 million), it is found in Europe, Asia, Africa, the Americas and Australia. During the winter it forms large roosts in reed beds, while in the nesting period its preferred habitat is agricultural areas. The nest is in the form of an open cup and made of mud and plant material.

燕子 【家燕】 一种雀形目类候鸟，其特征为叉状尾巴，锐弯曲线的翅膀和小而直的深灰色鸟嘴。这种鸟分布较广（世界上燕子的数量约为2亿只），在欧洲、亚洲、非洲、美洲和澳洲都比较普遍。在冬季，它们会用苇枝筑巢，而在鸟窝未建成之前，则喜欢在农耕区栖息。鸟窝为开杯式，是用泥土和植物材料做成的。

让它无法重新飞起来。这种燕子分布在欧洲、亚洲、非洲以及地中海南部大部分地区，尤其是有人类居住的地方，栖息地的海拔可达2 000m。它们倾向于在岩石、树木的天然空洞中筑巢，但是，大部分是在人造环境中，比如建筑的屋檐和水沟旁。

Swift [*Apus apus*] A migratory bird of the Apodiformes order, it is 15-18 cm long and has a wingspan of 35-40 cm. It is peculiar in that the femur is directly connected to the claw. This feature ensures that it never touches the ground throughout its life: if it perches on the ground, the limited functionality of the legs is such that it does not resume flight. It lives almost everywhere in Europe, most of Asia and Africa and the southern Mediterranean, especially in anthropic environments and up to 2,000 metres of altitude. It builds its nest in natural cavities in rocks or trees, but more often in artificial environments, such as gutters or the eaves of buildings.

楼燕 【楼燕】 候鸟，长15～18cm，翼展35～40cm。其股骨直接与爪子相连，这是非常罕见的。这种特点可以保证它们在整个生命期都不需要接触地面，如果它停落地面，其腿部有限的功能会

The Baron in the Trees Written by Italo Calvino in 1957, Il barone rampante (The Baron in the Trees) is a novel, the second chapter of the heraldic trilogy I nostri antenati (Our ancestors) which also includes Il visconte dimezzato (The Cloven Viscount) (1952) and Il cavaliere inesistente (The Nonexistent Knight) (1959). Set in Ombrosa, a fictional village in the Ligurian Riviera, it tells the story of a young baron, Cosimo Piovasco di Rondò, the eldest son of a lapsed noble family. After a futile argument with his father on June 15th 1767, the adolescent Cosimo climbs the trees in the garden of the house and never comes down. From that point on, his life is spent up in the trees until the day when, to everyone's surprise, Cosimo climbs to the top of a high tree, and from there grabs onto a balloon passing by, and disappears over the sea. The deep and extreme interaction between tree and man - Cosimo, who doesn't want to touch the earth "even when dead"

- takes on symbolic as well as poetic characters in the novel. Cosimo is a lonely man in his condition but he manages to interact in an incredibly wide and diversified way with the world on the ground, even so far as to create community. The Vertical Forest, a big tree for living, can be interpreted as an experiment that puts a range of conditions and suggestions in place imagined by Calvino's novel, interpreting the man-tree, tree-man interaction on a different level.

>> see big tree

《树上的男爵》由卡尔维诺于1957年所著。《树上的男爵》是一本小说，在第二章关于我家族祖先的描述中，提到了分成两半的子爵（1952）和不存在的骑士（1959）。故事从利古里亚海滨的一个虚构的村庄说起，这是一个关于一位年轻男爵柯西谟——一个没落贵族家庭的长子——的故事。1767年6月15日，在与父亲一次徒劳的争吵后，年少的柯西谟爬到家里花园的一颗树上，从此再也没有下来。从那时开始，他的一生都在树上度过。直到有一天，人们惊奇地发现，柯西谟爬到了一颗大树的顶部，紧紧抓住一个飞过的气球，消失在海面上。柯西谟，他"至死"都不想再触碰大地，这一树与人之间深入而极致的互动，极具象征性且颇有诗意地突显了小说中的人物。柯西谟在当时的条件下是很孤独的，但是，他却找到了一种博大且多样化的与地面上的世界互动的方式。垂直森林，一棵可以居住的大树，可以被解释为是根据卡尔维诺小说中的构想进行的一个实验性建设，它从不同层面上解释了人与树、树与人的互动。

>> 参见：大树

The house of the woods Among the reference projects for the Vertical Forest, mention should be made of "the house in the woods", built in 1969 by Cini Boeri, Stefano's mother. Built between Osmate and Taino and not far from Lake Maggiore in the province of Varese, the house was set completely within a birch forest. The layout is structured so as to avoid the felling of large trees. The different areas of the house (day, night, parents, children and guests) are differentiated through the joints of the building, generating more or less deep recesses that are home to plants. The outer wall is of exposed concrete, with many openings onto the outside and very few inside walls, while the rooms are divided by the use of sliding walls and different dimensions.

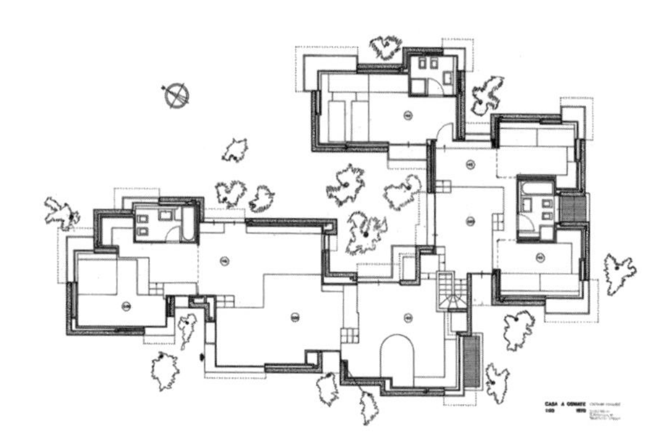

森林之屋 在垂直森林的参考项目当中，提到了由博埃里的母亲奇尼·博埃里设计的“树林之屋”。它位于奥斯马泰和泰诺之间，距离马焦雷湖不远的地方，一间完全在桦树林里的房子。房子的布局经过特别的设计以避免大树倒塌。房间的不同区域（日区，夜区，父母，儿童，宾客）通过不同的组合链接、区分开来，并创造了或多或少，为了树而退让出来的居所。建筑外墙是清水混凝土，有很多朝外的开口，很少内墙，而房间则是用滑行墙分隔成了不同大小。

Third Landscape The concept of "Third landscape", coined by the French landscape architect Gilles Clément, defines the sum of the areas abandoned by man in which nature has regained control, such as brown-field sites where brambles and weeds grow or traffic dividers where weeds grow: widespread residual and very different formations but crucial for the conservation of biodiversity. The absence of any human activity joins these fragments of landscape to the great places of the planet dominated by nature: institutionally protected uninhabited regions, protected areas, parks and reserves, naturally created reserves (inaccessible places, deserts, mountain peaks ...). According to Clément the Vertical Forest presents various similar aspects to the Third landscape, especially the variety of plant species present and the widespread sense of boundary between plant micro-landscape and anthropocized mi-cro-landscape. At the same time the French landscape architect identifies two elements of fundamental differentiation in the biodiversity conditions of the Vertical Forest and the Third landscape: in the former they are the result of an initial act of design and are arrived at very quickly; in the latter they are completely spontaneous and require several years.

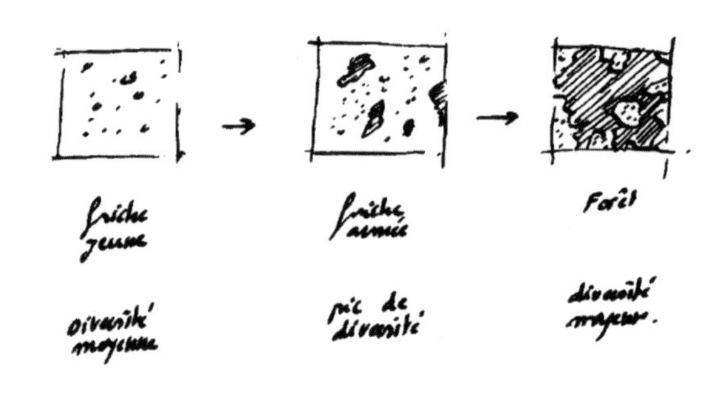

第三景观 “第三景观”的概念是由法国景观建筑师吉尔斯·克莱门特提出的，指的是被自然重新掌控，但被人类所冷落的地区总称，比如荆棘和野草生长的棕地，或野草生长的交通隔离带。这类地区有很多垃圾，有不同的形态，但是对于维护生物多样性却非常重要。人类活动严重影响了我们这个由自然主导的星球，需要制度化地保护那些无人居住的地区、保护预留区（难靠近的地方，荒漠和山峰）和公园等。根据克莱门特的观点，垂直森林与第三景观存在很多共同点，尤其是在植物微景观和人类中心主义微景观的界定意义、以及植物物种的多样性方面。同时，这位法国景观建筑师定义出垂直森林和第三景观在生物多样性方面的两个根本区别：前者是基于最初的设计，并产生得很快；后者则完全是经过多年时间自发形成的。

Three-dimensional facade Contrary to what happens in traditional buildings, the facades of the Vertical Forest are not conceived as two dimensional elements acting purely to separate the inside from the outside, but rather as a complex system with three dimensions, where height and width are combined with depth in terms of development for the plants. In the first created Vertical Forest the real architectural facade consists of a wall of hollow bricks, onto which an insulating layer of thermal fiber is superimposed and an outer coating formed by a ventilated wall of porcelain panels supported by metal uprights. The outermost layer of the facade is made up of the vegetation, which acts as a natural filter and produces beneficial microclimatic conditions indoors.
>> see balconies, microclimates, colours, materials

三维立面 与传统建筑立面截然相反，垂直森林的立面并不是单纯地将室内室外分开的二维元素，而是一个三维的复杂体系，其高度和宽度，都是与植物生长深度结合的。现实建筑立面上的垂直森林是由空心砖的墙面构成，上面叠加了带保暖纤维的绝热层，然后外涂层是由金属支撑的，陶瓷板之后还能通风。立面的最外层是由植物构成，这些植物被当成天然的过滤器，还可以为室内提供有益的小气候环境。
>> 参见：阳台，小气候，色彩，材料

Urban forest The first Vertical Forest built in Milan is home to about 21,000 plants, belonging to more than 100 different plant species: the equivalent of a 2 hectares forest grafted onto a small piece of land of 1,500 m^2 in the centre of a big city.
>> see biodiversity, plant species, humans

城市森林 首个垂直森林在米兰建立，它是21 000棵植株、100多个不同植物品种的所在地，这相当于将2hm^2的森林，移植到了一个在大城市中心面积仅为1500m^2的土地上。
>> 参见：生物多样性，植物品种，人类

Urban sensor The Vertical Forest is designed to become an urban sensor for the spontaneous recolonization of plants and animals in the city, especially for seeds carried by the wind and the movement of flying insects. On the roofs of the Vertical Forest free levels can be created for the colonization of nomadic species. A widespread system of more Vertical forests would support monitoring of this evolutionary development. >> see biodiversity, continuous city, cultural cell, replicability, laboratory-roof

城市传感器 垂直森林被设计成了一个城市里的感应器，可以监测到动植物的再迁徙，尤其是那些被风和昆虫无意识携带的种子。在垂直森林的顶部有一个自由平台，这里可以作为游牧物种的聚居地。垂直森林有一个广泛分布的探测系统，帮助我们监测这种演变的发展过程。

>> 参见：生物多样性，连绵城市，文化细胞，可复制性，实验室屋顶

Urban traffic (Anti-congestion of) If built near public transport hubs, the Vertical Forest can help reduce urban traffic both entering and leaving the city, and encouraging the return to the city centre of families and individual commuters who have moved their homes into suburban areas.

>> see continuous city, anti-sprawl device, replicability

城市交通（反拥挤） 如果垂直森林建在临近公共交通枢纽的区域，则会利于减少进入和离开城市的城市交通量，它鼓励那些已把家搬到近郊区的家庭和个人通勤者回到市区。

>> 参见：连绵城市，反城市蔓延的设备，可复制性

Villas (superimposition of) One tower of the Vertical Forest is equivalent to the superposition of a series of villas with gardens. Such a model calls to mind the famous experimental Modernism projects: first of all the immeubles villas designed by Le Corbusier in 1922 and later repeatedly reinterpreted by other designers. Compared to previous similar projects, the Vertical Forest differs in two main ways: its application as an already built environment (the immeubles villas and their subsequent variations were paper experiments) and the conceptual reversal, which focuses on the plant component rather than the architectural gesture.

>> see anti-anticity, mute architecture, anti-sprawl device

别墅（叠加） 垂直森林一栋大楼就相当于叠加了一系列带花园的别墅。这一模型让人想起了著名的现代实验性项目：1922 年由勒 · 柯布西耶设计的别墅－公寓。之后又有其他设计师对此进行重新演绎。与之前类似的项目相比，垂直森林主要在两个方面存在不同：一方面，它是已建成的建筑，而别墅－公寓及其后来的变体都是书本实验；另一方面则是概念的逆转，它关注的是植物而非建筑造型。

>> 参见：反－反城市，沉默建筑，反城市蔓延的设备

White Wagtail [*Motacilla Alba*] A passerine member of the Motacillidae family with the size of a common sparrow and found throughout Europe and Asia. It prefers habitats close to water even if it has a remarkable ability to adapt to different environments. It needs open spaces with close-cropped herbaceous vegetation, alternating with areas of bare soil. It adapts very well to man-made spaces and unlike many other birds, it is not uncommon to see it even in mountainous areas. gutters or the eaves of buildings.

白面鸟 【白鹡鸰】 鹡鸰科中雀形目的一员，体型与普通麻雀相似，在整个欧洲和亚洲地区都有繁殖。虽然其环境适应能力很强，但偏好栖息于靠近水的地方。它们需要在修剪较短的草本植被与裸露土壤相间的公共空间生存。它们可以较快地适应人造空间，并且与其他鸟类不同，它们在山区也很常见。

Wind tunnel The wind was the main climate factor considered in developing the first Vertical Forest: on the one hand for an assessment of its direct impact on vegetation and on the other hand due to the fact that the weight and size of the trees when subjected to the power of the wind are able to transmit a series of intense and complex forces to the structure of the buildings such as to require extremely careful verification. For this purpose, in the preliminary planning stage an initial experimental verification was performed through a test conducted in the wind tunnel at the Milan Polytechnic. The tests made it possible to determine the geography of stress on the trees and the building structure, highlighting local phenomena that increased speed.
The study was conducted using a 1: 100 scale model, with trees located on spring balances to evaluate aerodynamic forces, moments and coefficients, taking into account the changes that the morphology of the city, both current and future, actually determines and will determine regarding the flow of wind in this context. The results obtained allowed an assessment of the sizing of the trees and the optimization of the distribution of the same on the facades of the towers. In a second project phase, a carefully calculated real-time model of the pot/tree/anchors/substrate package was subjected to a genuine test carried out in the wind tunnel at Florida International University in Miami. The characteristics of the preselected structure allowed testing of the aerodynamic coefficient of real trees, with windy conditions up to 190 km/hour. The plants were subjected to test speeds up to 67 m/s, with measurements of forces and moments at the base.
Thanks to such intensive tests it was possible to determine the characteristics and intensity of the reactions of the trees affected by the wind and of the forces transmitted to the building structures through the turf and the roots system.
>> see plants-selection, structure

风洞 在首个垂直森林的开发过程中，风是被主要考虑的气候因素：一方面是因为风会对植物产生直接影响；另一方面，当树的大小和重量承受强风时，就会向建筑结构传送一系列巨大和复杂的力，

需要特别加以确认校核。为此，在最初的规划阶段就于米兰理工大学的风洞里先进行了一项验证测试。这项测试确定出树木和建筑结构的应力分布，还特别指出会发生风速增加的局部现象。

这项研究使用了一个 1：100 的模型。树木被放置在弹簧秤上，以对空气动力、力矩和系数等进行评估。同时，还考虑到了当前和未来的建成城市形态，因为城市形态会对这种环境下风的流量起到决定性作用。所获得的结果被用于帮助评估和优化大楼立面上植物的规格和布局。项目二期时，在迈阿密佛罗里达国际大学风洞中，又对种植盆、植物、锚、封装基板的实体模型进行了仔细计算与真实测试。研究者使用预先选择的结构测试了真树在风速 190km/h 条件下的空气动力系数；在 67m/s 的风速下，测试了植物底部的力和力矩。

全靠如此高强度的测试，才使植物在受到风速影响下的特性以及反应强度得以明确，同时还确定了草皮和根系传给建筑结构的力的特性。

>> 参见：植物配置，结构

Wood Pigeon [*Columba palumbus*] A widespread species of pigeons, present in the Middle East and all of Europe, except for Iceland and northern Scandinavia. It is 38-40 cm long, and its favoured habitat is forests of all kinds, especially the outer edges, but also gardens and city parks. The nest consists of straw and branches and is built especially among the higher branches of trees.

斑尾林鸽【斑尾林鸽】 一种较为常见的鸽子品种，原产于中东和除了冰岛与斯堪的纳维亚北部以外的所有欧洲地区。它们体长 38～40cm，偏爱栖息于各种类型的森林，特别是森林的边缘地区，此外，在花园和城市公园中也很常见。林鸽常用稻草和树枝筑巢，特别喜欢在树木的高枝上筑巢。

LEARNING FROM VF01

向垂直森林 01 号学习

1

What is VF01 Learning from Vertical Forest 01: 10 reasons
什么是垂直森林01号 向垂直森林01号学习的十个理由

1 VF01 is a project for the environmental survival of contemporary cities.
垂直森林01号是当代城市中的环境生存项目。

2 VF01 multiplies the number of trees in cities.
垂直森林01号倍增了城市中的树木。

3 VF01 is a tower for trees inhabited by humans.
垂直森林01号是树木与人类共同居住的塔楼。

4 VF01 is an anti-sprawl device.
垂直森林01号是反城市蔓延的设备。

5 VF01 demineralizes urban surfaces.
垂直森林01号净化了城市的矿物表面。

6 VF01 reduces the pollution of the urban environment.
垂直森林01号降低了城市环境中的污染。

7 VF01 reduces energy consumption.
垂直森林01号减少了能源消耗。

8 VF01 is a multiplier of urban biodiversity.
垂直森林01号是城市生物多样性的放大器。

9 VF01 is an ever changing urban landmark.
垂直森林01号是一个不断变化的城市地标。

10 VF01 is a living ecosystem.
垂直森林01号是一个活的生态系统。

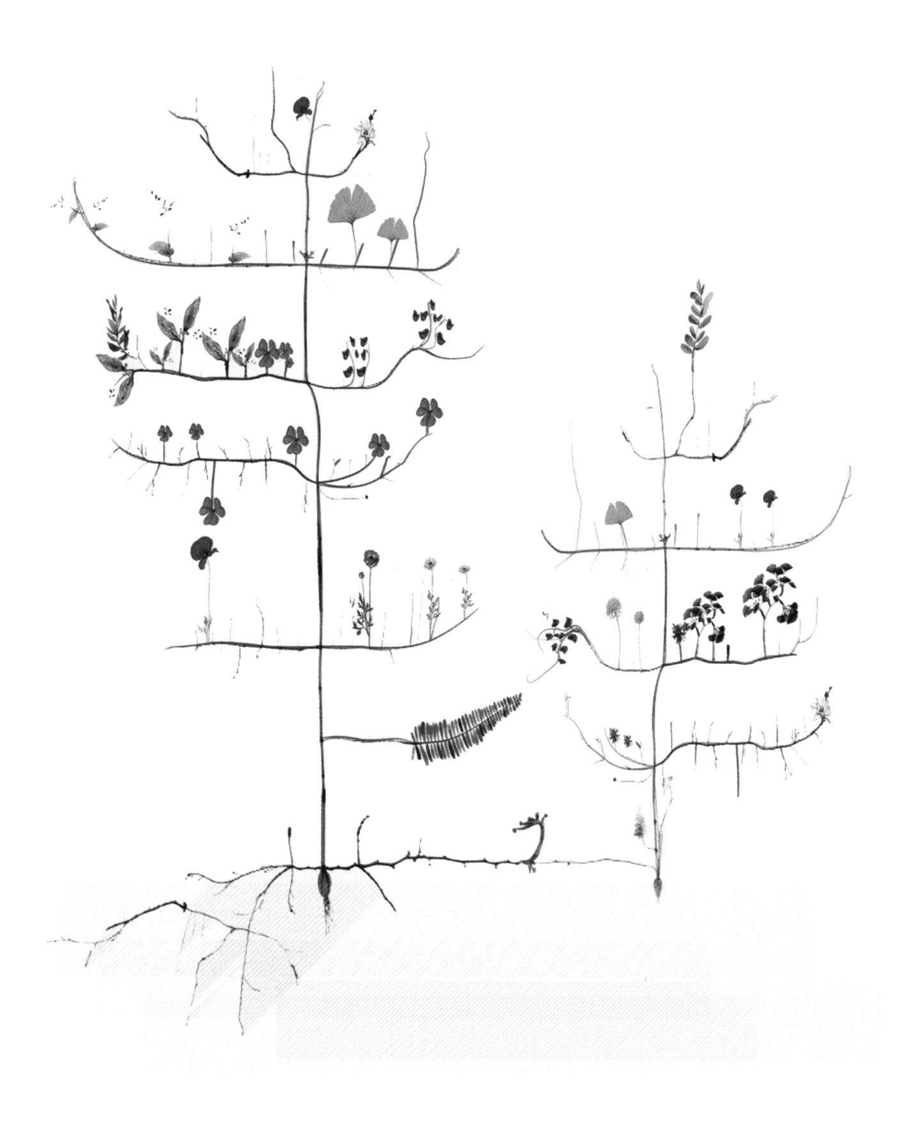

1.1

VF01 is a project for the environmental survival of contemporary cities.

VF01 is a new generation of high-rise urban buildings completely covered by the leaves of trees and plants.
VF01 is an architectural device that promotes the coexistence of architecture and nature in urban areas, and the creation of complex urban ecosystems.

垂直森林01号是当代城市中的环境生存项目。

垂直森林01号是城市新一代高层建筑，它完全被树木和植物覆盖。

垂直森林01号是推动城市中的建筑与自然共生的仪器，它是复杂的城市生态系统的创新。

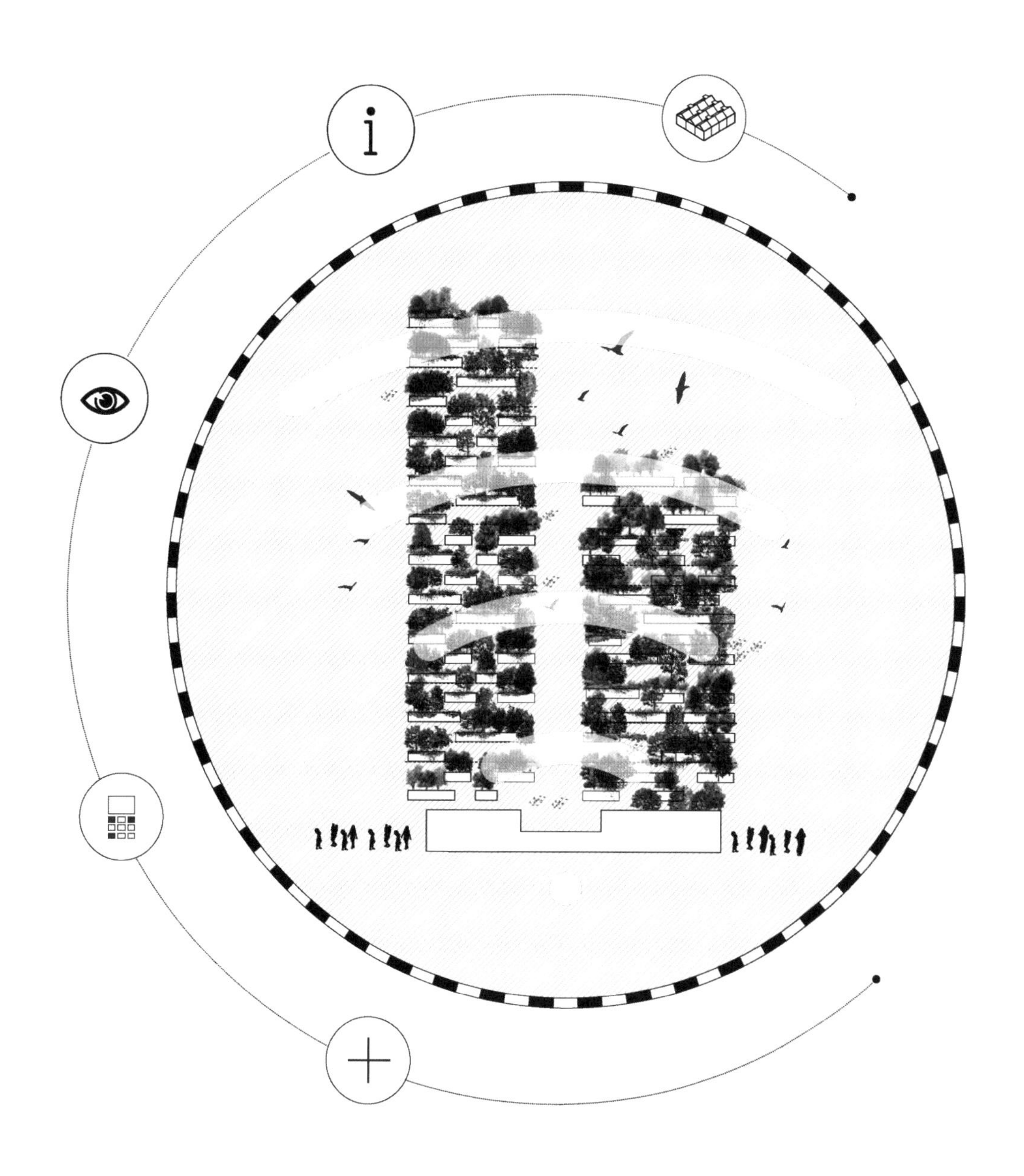
i

1.2

VF01 multiplies the number of trees in cities.

VF01 grafts the equivalent of thousands of square meters of forest and undergrowth onto a few hundred square meters of urban space. The terraces, genuine outdoor extensions of the living spaces are home
to over a dozen trees (ranging from 3 to 9 metres in height), numerous shrubs and flowering plants.
If about 350 trees make up a 1 hectare forest, the over 700 trees of VF01 are equivalent to 2 hectares of woodland and ground level undergrowth.

垂直森林01号倍增了城市中的树木。

垂直森林01号于几百平方米的城市空间之中移植了相当于数千平方米的森林和丛林。露台，是真正的居住空间的户外延伸，也是众多树木（高达3～9m）、不计其数的灌木以及花卉的家。

如果350棵树可以造就1hm^{2}的森林，那么种植超过700棵树的垂直森林01号的绿量就相当于2hm^{2}的森林和林下植被。

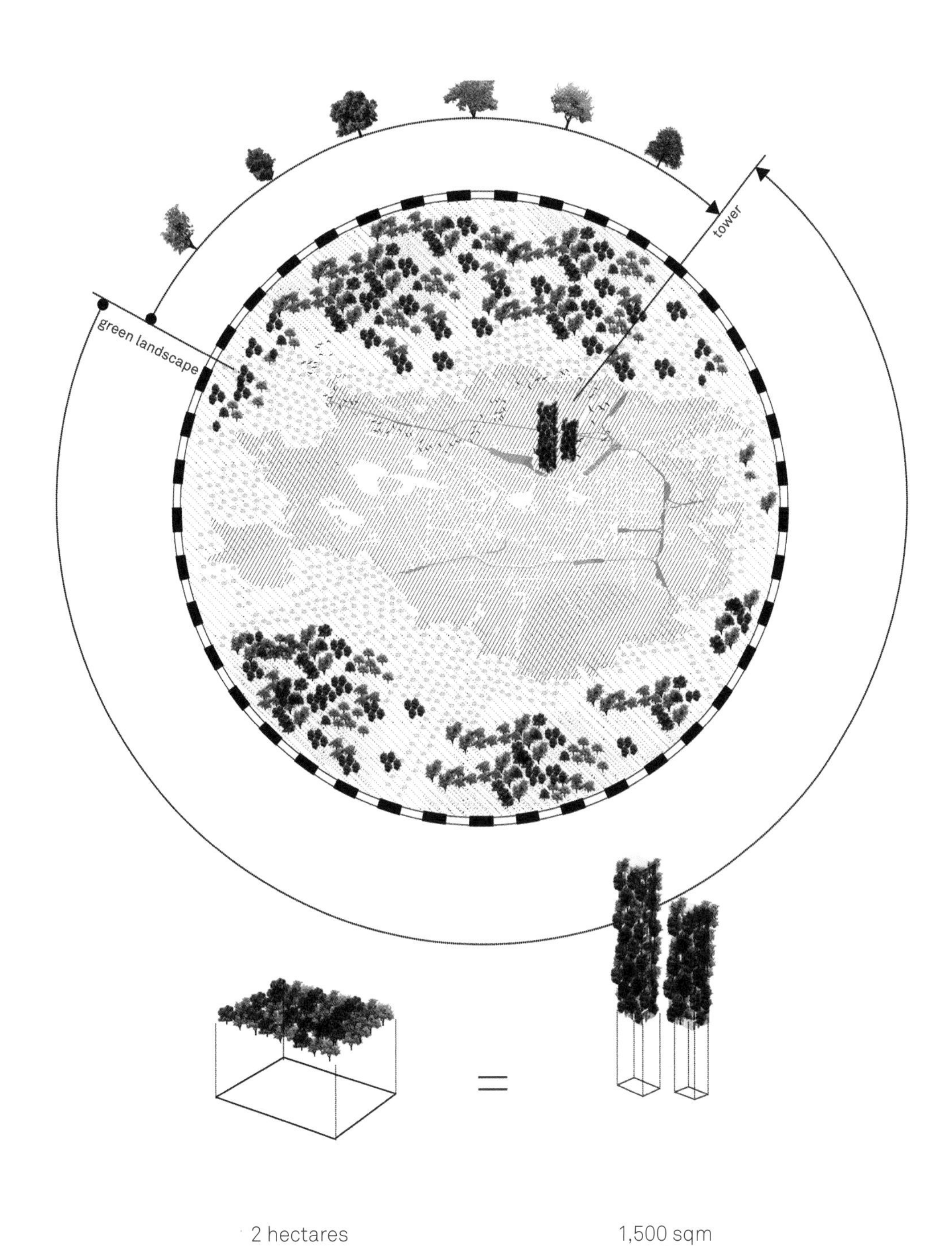

2 hectares
2 hm^2

1,500 sqm
1 500 m^2

1.3

VF01 is a tower for trees inhabited by humans.

VF01 sets the living standard between humans and trees within the built environment, establishing an amount of 2 trees, 8 shrubs and 40 bushes for each human being.
VF01 houses about 1.7 km of pots that are 1.10 metres high and up to 1.10 metres wide.
The soil contained in the pots is a mix of agricultural soil, organic matter and volcanic material that allows the reduction of the volume weighting on the perimeter of the balconies.
Such diversity and typology of plant species within the urban centre works as a point of reference and a tool for urban policies directed to the inclusion of plant and animal species inside the man-made urban context, promoting the development of different urban biodiversity dissemination sites.

垂直森林01号是树木与人类共同居住的塔楼。

垂直森林01号在建成的环境中设置了人与树木之间的生存标准，它为每一个人都对应种植了2棵大乔木、8棵大灌木和40棵小灌木。

垂直森林01号容纳了近1.7km长的种植盆（高1.1m，宽1.1m）。

培养盆中的土壤混合了农业用土、有机填料以及火山质，可以减轻阳台周边的荷载。

在城市中心运作如此多样和丰富的植物类型，垂直森林可以作为城市政策的参照和工具，直接推动了植物和动物进入到人造的城市文脉中，促进了不同地区的城市生物多样性传播。

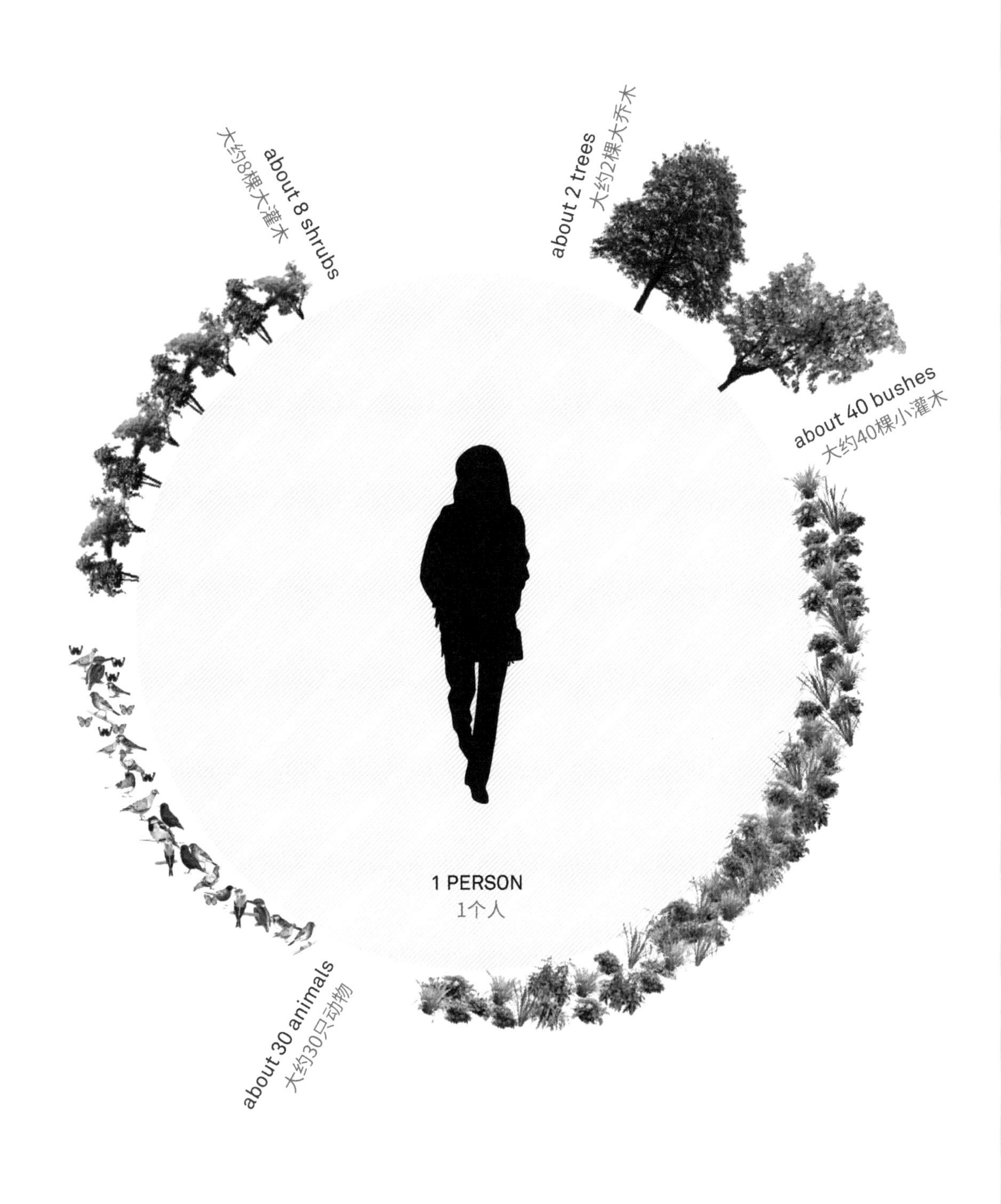
about 8 shrubs
大约8棵大灌木
about 2 trees
大约2棵大乔木
about 40 bushes
大约40棵小灌木
1 PERSON
1个人
about 30 animals
大约30只动物

1.4

VF01 is an anti-sprawl device.

VF01 constitutes an alternative urban environment that allows to live close to trees, shrubs and plants within the city; such condition can be generally found only in the suburban houses with gardens, which are a development model that consume agricultural soil and which is being now recognized as energy-consuming, expensive and far from communal services found in the compact city.
Through densifying the urban fabric, VF01 creates new and innovative relationships of proximity between nature and the built environment, creating new landscapes and new skylines.

垂直森林01号是反城市蔓延的设备。

垂直森林01号建构了另一种城市环境，在此，城市生活也可以贴近乔木、灌木以及其他植物。这种生活条件常常发生在郊区的住宅与花园中，但是，现在人们已经意识到这是一种消耗农用地的发展模式，不仅消耗能源，代价昂贵，而且远离城市公共服务。

通过城市结构的致密化，垂直森林01号创造了一种自然和城市环境间的亲密关系，创造了新的景观和天际线。

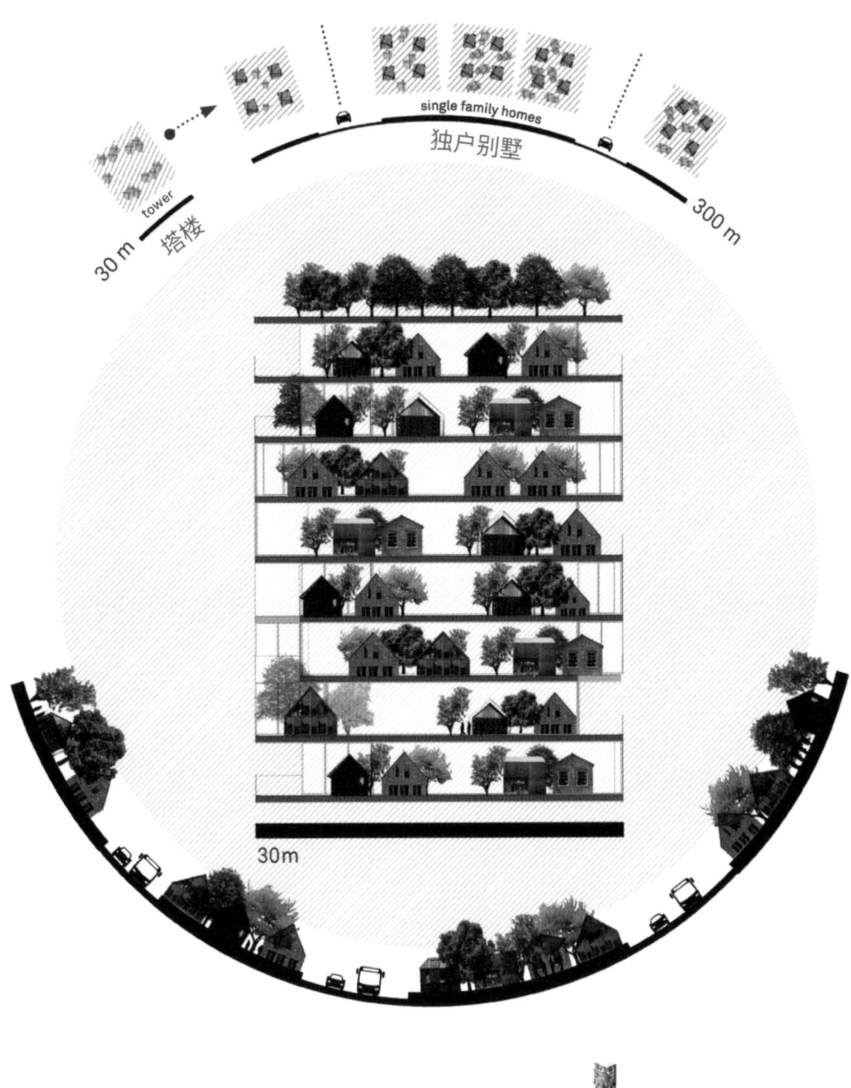

50 hectares
50 hm²

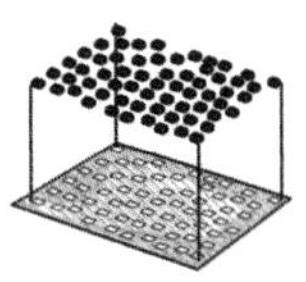

=

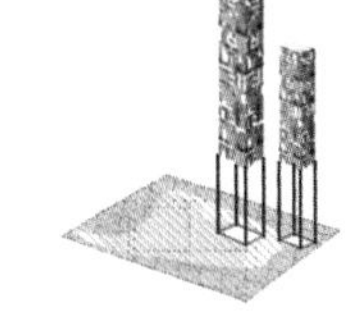

1,500 sqm
1 500 m²

1.5

VF01 demineralizes urban surfaces.

VF01 is a high-density forestation project that increases green and permeable surfaces in the city and reduces the heat island caused also by sunlight reflected from glass façades.
Together with Green Roofs, Vegetable Gardens, and Vertical Gardens, VF01 belongs to a new generation of environmental regeneration projects aimed at improving the quality and variety of everyday life in contemporary cities.

垂直森林01号净化了城市的矿物表面。

垂直森林01号是高度森林化的项目，即增加了城市中的绿化和渗透面，同时减少了由玻璃立面反射日光造成的热岛效应。

与绿色屋顶、蔬菜园以及垂直花园一样，垂直森林01号属于新一代的环境再生项目，旨在改善当代城市中每天生活的质量和多样性。

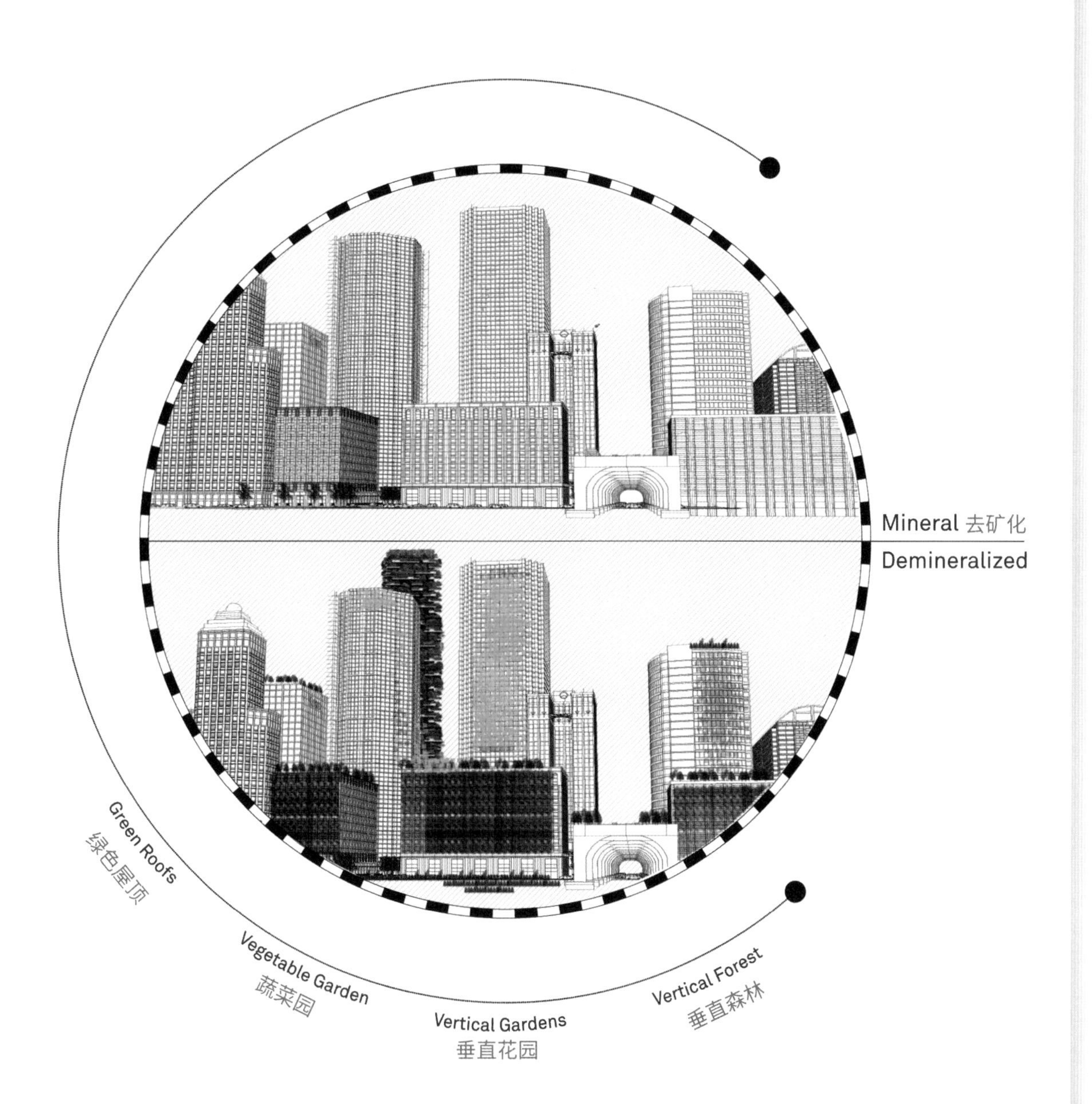
Mineral 去矿化
Demineralized
Green Roofs
绿色屋顶
Vegetable Garden
蔬菜园
Vertical Gardens
垂直花园
Vertical Forest
垂直森林

1.6

VF01 reduces the pollution of the urban environment.

The vegetation within VF01 is designed in such a way as to form a continuous green filter between the inside and the outside of the inhabited areas, able to absorb the fine particles produced by urban traffic, to produce oxygen, to absorb CO2 and to shield the balconies and interiors from noise pollution. The benefits resulting from a reduction in pollution are apparent not just for the building itself and its residents, they also contribute to improving air quality in the whole city.

垂直森林01号降低了城市环境中的污染。

垂直森林01号中的植物被设计定义为持续的绿色净化器，安置在居住区的室内与室外之间，能够吸附城市交通产生的细小颗粒，产生氧气，吸收二氧化碳，并在阳台上制造一层隔离噪声与污染的屏障。这些因减少环境污染而带来的显著效益，不仅是有益于建筑本身与住户的，更为整个城市空气质量的提高作出了贡献。

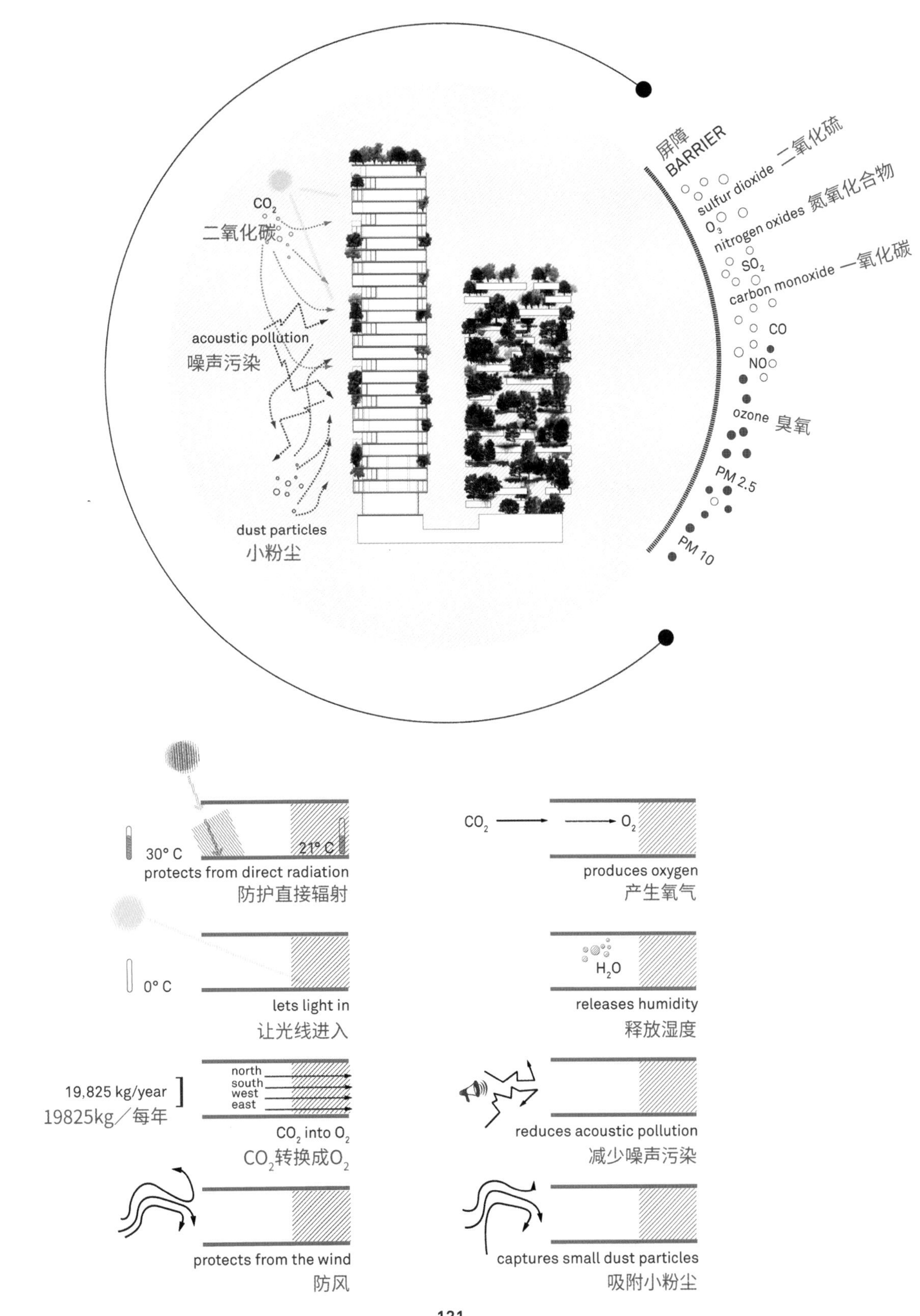

CO_2
二氧化碳
acoustic pollution
噪声污染
dust particles
小粉尘
屏障
BARRIER
sulfur dioxide 二氧化硫
O_3
nitrogen oxides 氮氧化合物
SO_2
carbon monoxide 一氧化碳
CO
NO
ozone 臭氧
PM 2.5
PM 10
30° C
21° C
protects from direct radiation
防护直接辐射
CO_2
O_2
produces oxygen
产生氧气
0° C
lets light in
让光线进入
H_2O
releases humidity
释放湿度
19,825 kg/year
19825kg／每年
north
south
west
east
CO_2 into O_2
CO_2转换成O_2
reduces acoustic pollution
减少噪声污染
protects from the wind
防风
captures small dust particles
吸附小粉尘

1.7

VF01 reduces energy consumption.

Trees and shrubs in the VF01 are irrigated with groundwater pulled by a pump system powered by solar panels located on the roof to all the pots in the building. The water used by the trees and shrubs returns purified in the atmospherein the form of water vapour. The process extracts heat from the surrounding environment.
The vegetation acts as a filter on the VF01 balconies determining a reduction of nearly 3 degrees between outside and inside temperature and – in summer – a decrease in the heating of the façades by up to 30 degrees.

垂直森林01号减少了能源消耗。

垂直森林01号屋顶的太阳能板给水泵供电，水泵可抽取地表水，灌溉垂直森林01号内每个种植池中的乔木和灌木。乔木与灌木将灌溉水转化成水蒸气净化空气。这个过程会从周围环境吸收热量。垂直森林01号阳台上的植物作为过滤器，可以保证室内温度比室外减少将近3°C，尤其在夏季，可以缓解建筑外表皮高达30°C的高温。

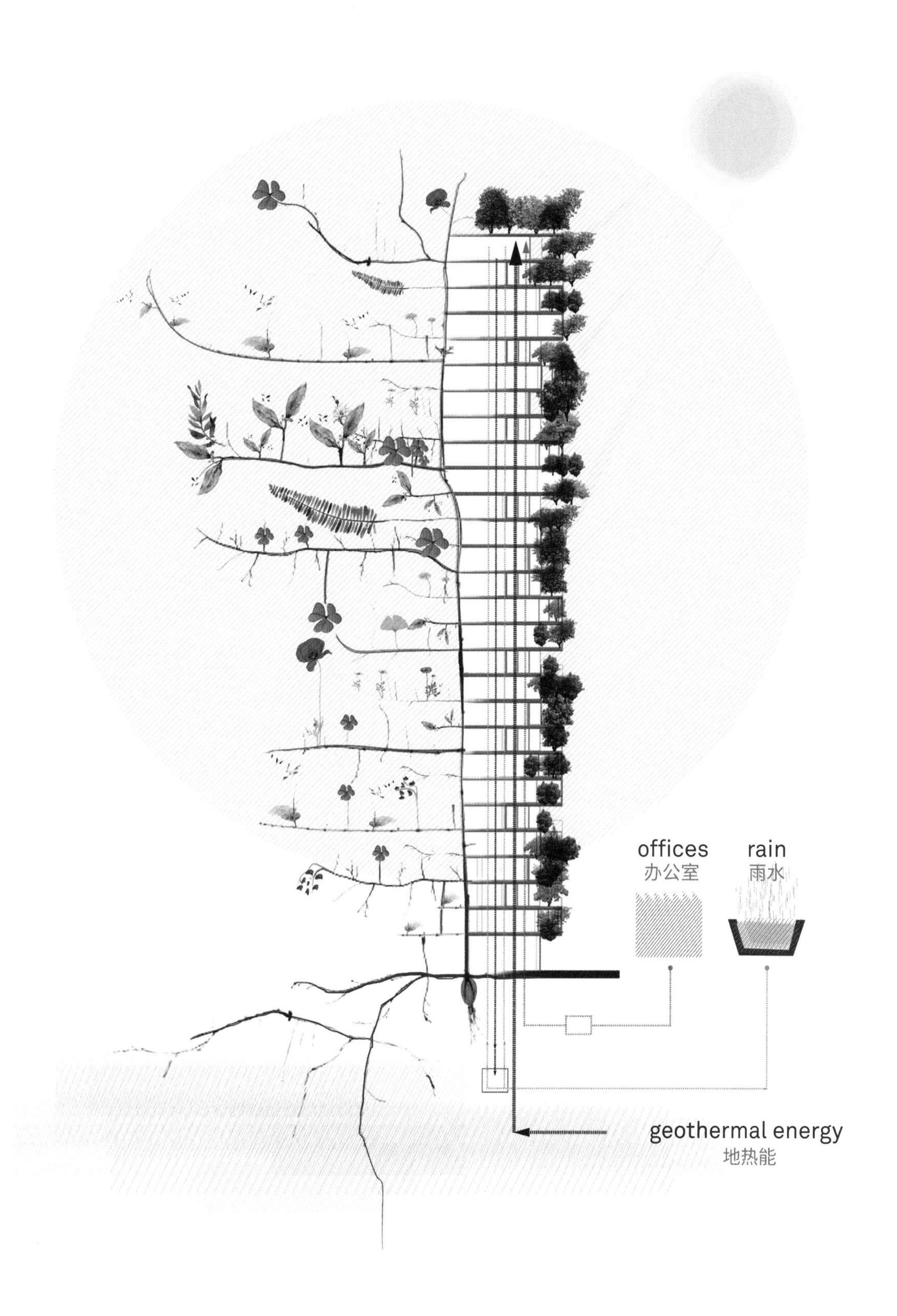

offices
办公室
rain
雨水
geothermal energy
地热能

1.8

VF01 is a multiplier of urban biodiversity.

VF01 hosts about 100 different plant species, including 15 species of trees, 45 shrubs and 34 types of perennials.
More than 20 species of birds have nested on the trees and bushes found in VF01, such as the martin, the redstart and pale swift.
Different insect populations live in the VF01 vegetation, some of which, such as ladybirds, were released inside the vegetation in order to fight plant pests without using pesticides.

垂直森林01号是城市生物多样性的放大器。

垂直森林01号容纳了约100种不同的植物品种，包括15种乔木、45种灌木和34种多年生植物。

超过20种鸟在垂直森林01号内的树木和草丛中筑巢，例如岩燕、红尾鸲和白腰雨燕。

不同的昆虫种群生活在垂直森林01号的植物中，其中一些，如瓢虫，生长于植物中，用于对抗植物虫害，从而避免了农药的使用。

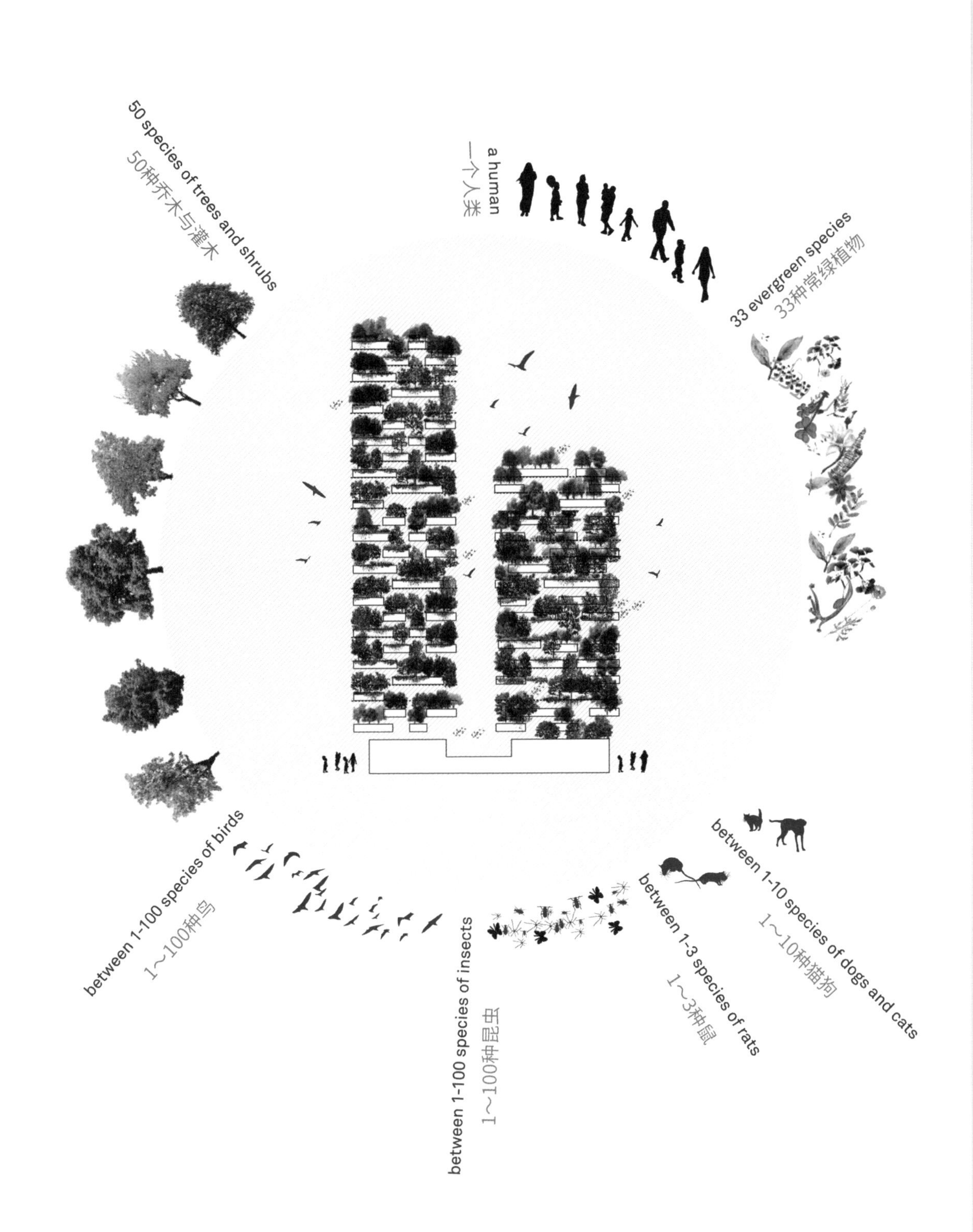
50 species of trees and shrubs
50种乔木与灌木
a human
一个人类
33 evergreen species
33种常绿植物
between 1-100 species of birds
1～100种鸟
between 1-100 species of insects
1～100种昆虫
between 1-3 species of rats
1～3种鼠
between 1-10 species of dogs and cats
1～10种猫狗

1.9

VF01 is an ever changing urban landmark.

Because of the variety of plant species hosted in the balconies and the presence (especially on the north walls) of many different deciduous trees, VF01 changes its skin and the colour composition of its living façades, according to the season variability and weather conditions.
Like the trunk of a tree, its outer shell turns it into a living urban archive, a witness to the slow and gradual growth of a new and rich urban ecosystem in the heart of the city.

垂直森林01号是一个不断变化的城市地标。

因为阳台中有不同的植物种类，尤其在北墙有许多不同的落叶乔木，垂直森林01号一直改变着它的外表和颜色，其鲜活的外墙，随季节与天气的变化而变幻。

就如同树干一样，它的外表使它变成了一个活生生的城市档案，见证了丰富的生态系统在城市中心缓慢而持续地生长。

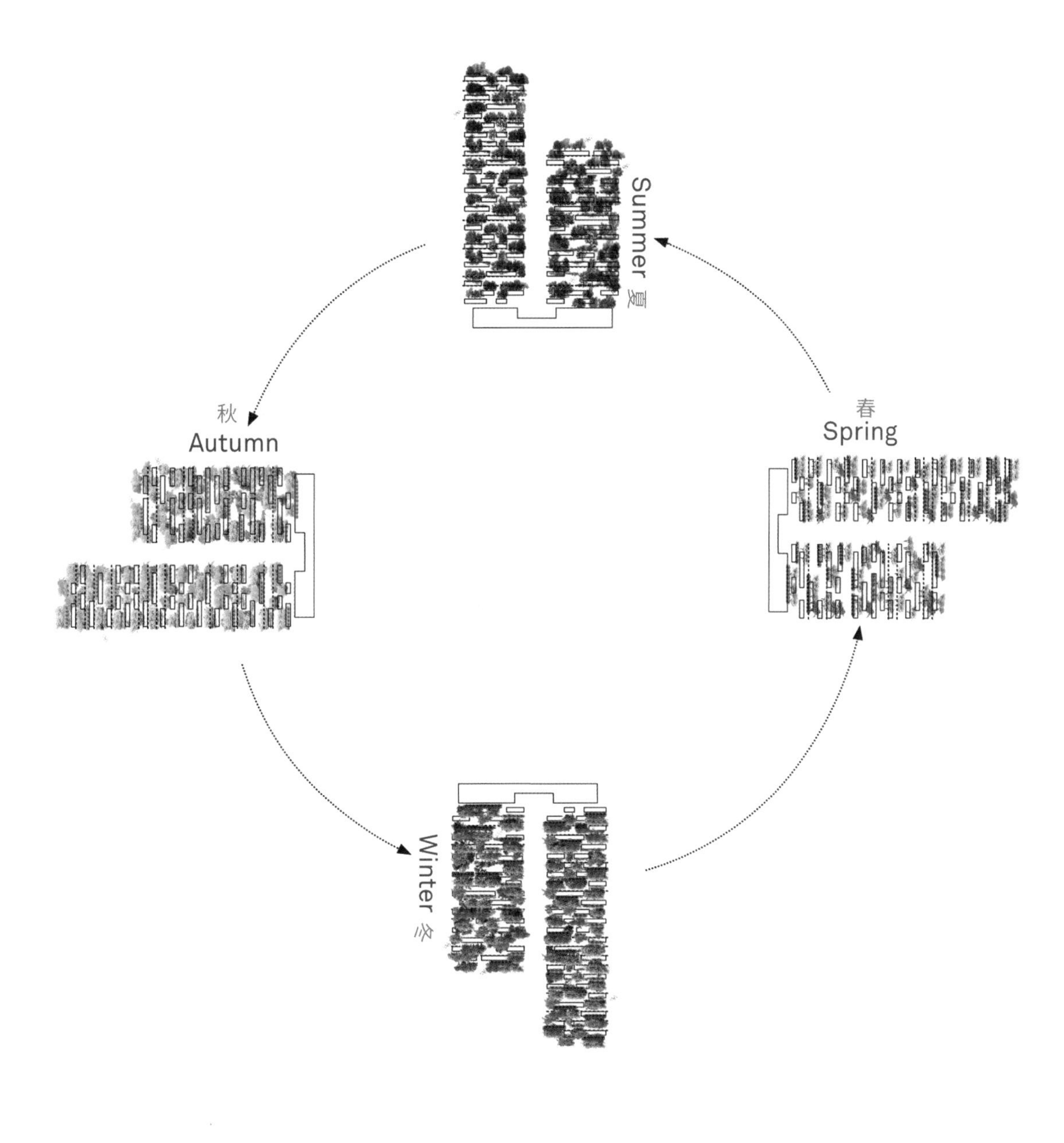
Summer 夏
秋
Autumn
春
Spring
Winter 冬

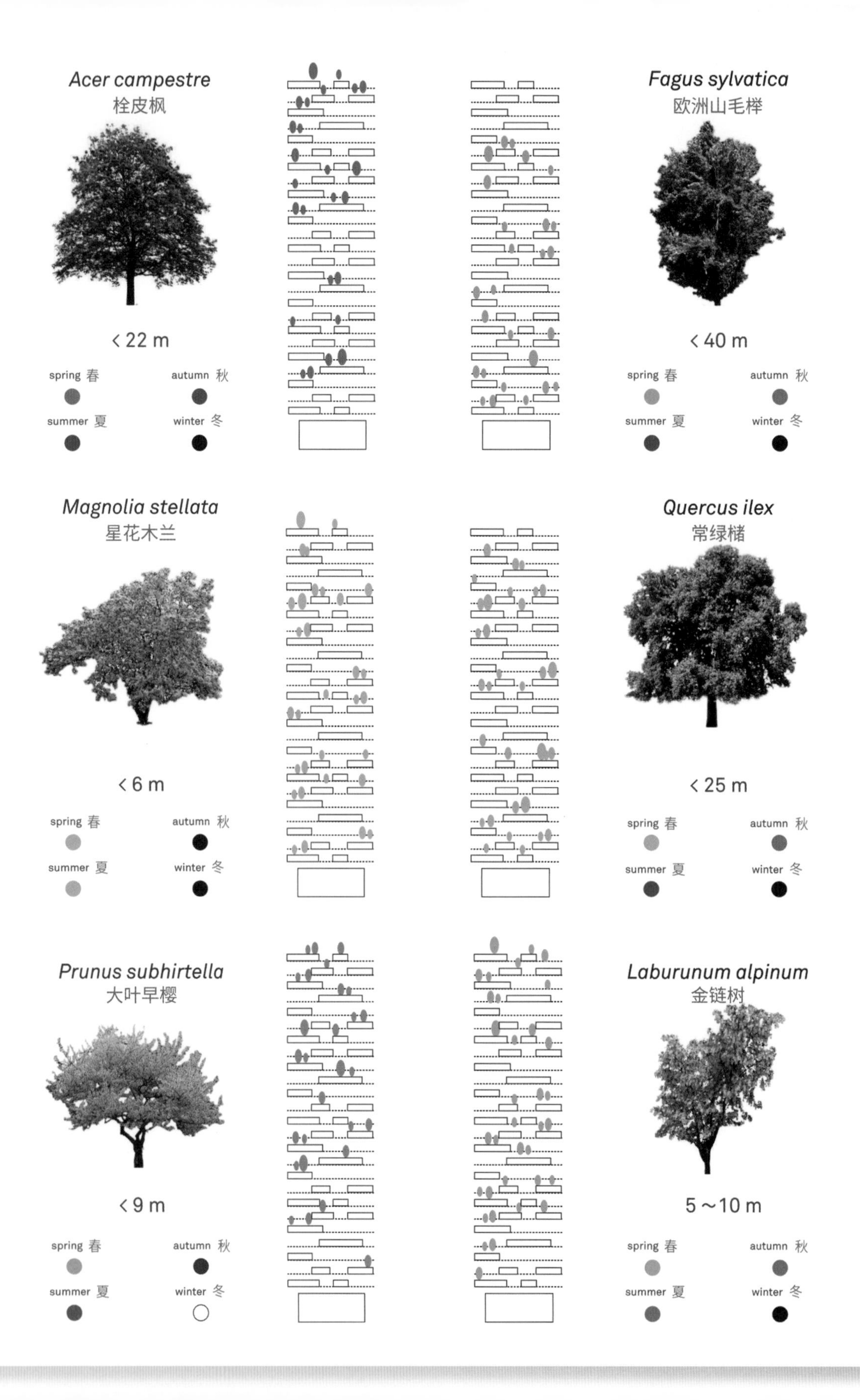
Acer campestre
栓皮枫
< 22 m
spring 春
autumn 秋
summer 夏
winter 冬
Fagus sylvatica
欧洲山毛榉
< 40 m
spring 春
autumn 秋
summer 夏
winter 冬
Magnolia stellata
星花木兰
< 6 m
spring 春
autumn 秋
summer 夏
winter 冬
Quercus ilex
常绿槠
< 25 m
spring 春
autumn 秋
summer 夏
winter 冬
Prunus subhirtella
大叶早樱
< 9 m
spring 春
autumn 秋
summer 夏
winter 冬
Laburunum alpinum
金链树
5～10 m
spring 春
autumn 秋
summer 夏
winter 冬

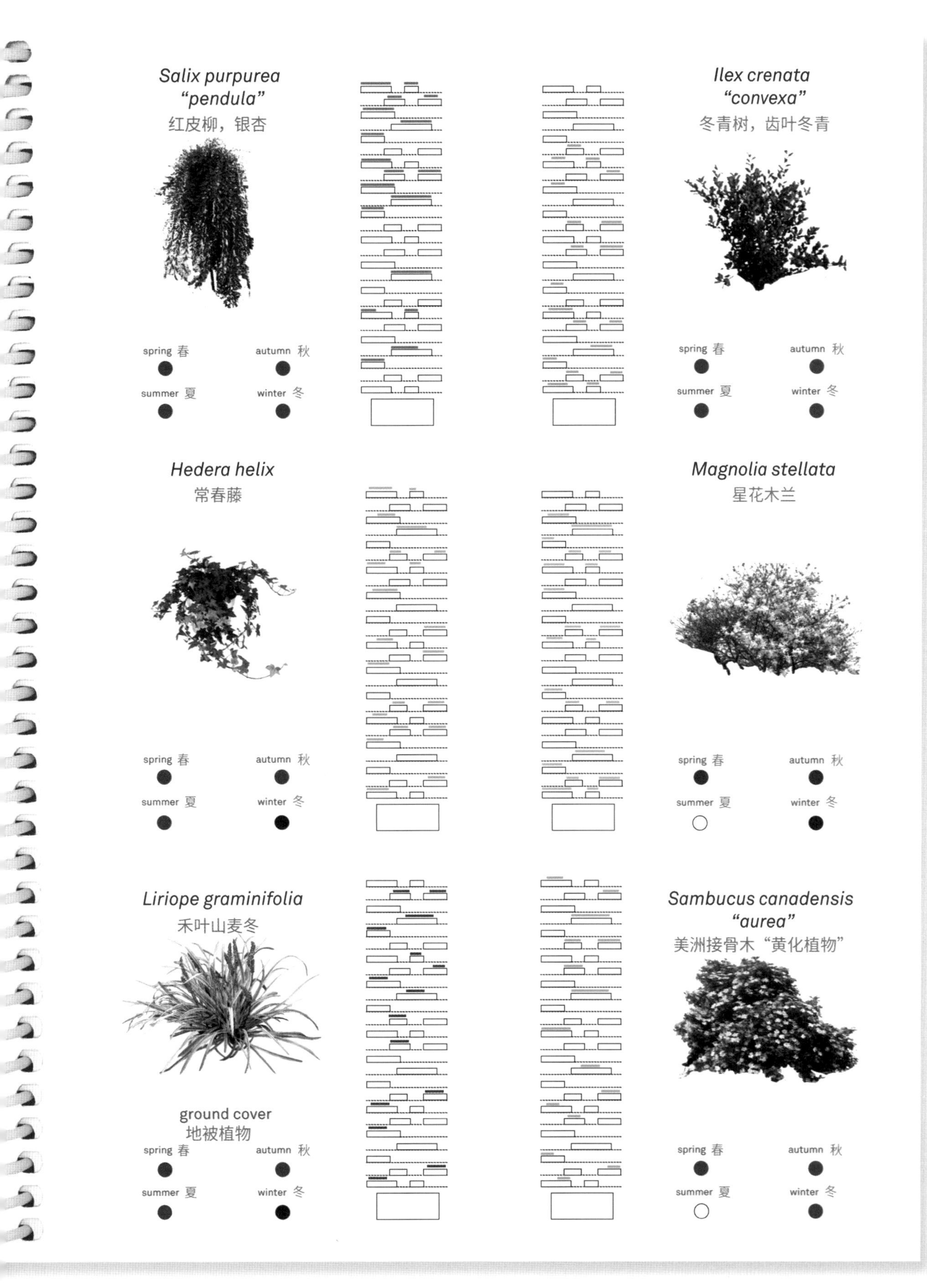
Salix purpurea
"pendula"
红皮柳，银杏
spring 春
autumn 秋
summer 夏
winter 冬
Ilex crenata
"convexa"
冬青树，齿叶冬青
spring 春
autumn 秋
summer 夏
winter 冬
Hedera helix
常春藤
spring 春
autumn 秋
summer 夏
winter 冬
Magnolia stellata
星花木兰
spring 春
autumn 秋
summer 夏
winter 冬
Liriope graminifolia
禾叶山麦冬
ground cover
地被植物
spring 春
autumn 秋
summer 夏
winter 冬
Sambucus canadensis
"aurea"
美洲接骨木"黄化植物"
spring 春
autumn 秋
summer 夏
winter 冬

1.10

VF01 is a living ecosystem.

The contact between the leaves belonging to trees and the ones belonging to shrubs and groundcovers planted at different levels transforms the façades of VF01 into potential vertical ecological corridors. As an ecosystem in a state of perennial transformation, VF01 needs a high professional centralized tree care maintenance.

Every 2-3 months, VF01 gardeners from inside the balconies take care of plants and once a year treeclimber arborists intervene to prune the trees from the outside. The health of trees and shrubs hosted in the towers of VF01 is monitored in real time by a centralized network of sensors located in every pot.

垂直森林01号是一个活的生态系统。

位于不同高度的乔木、灌木和地被植物的树叶将垂直森林的立面变幻成了潜在的垂直生态走廊。作为一个常年运转的生态系统，垂直森林01号需要高水准且专业化的集中式的植物养护管理。

每隔2~3个月，垂直森林的园丁们会从阳台内照看这些植物，同时每年，树木栽培师都像树的攀行者一样，从外部对树木进行一次修剪。通过种植盆内的传感器，垂直森林01号内的乔木和灌木的健康状况都在中央网络的实时监控之下。

1-2 / year　1~2次／年
cutting external parts of trees
修剪植物的外侧

4-6 / year　4~6次／年
cutting internal part of trees
修剪植物的内侧

6 years　每6年
control and diagnostics
植物病害检查

2

Towards a Forest City
迈向森林城市

The Forest City is the prototype for a new style of life that develops directly from the extension, multiplication and evolution of the Vertical Forest.
The Forest City is a vertical city, that brings together plant life normally found spread over hundreds of hectares of forest into a space of one or two square kilometres of urban surface. The Forest City is a settlement for humans, trees, plants and other animal species based around a minimum standard of 2 trees per inhabitant.
The Forest City is a complex and integrated ecosystem, which combines environmental sustainability (anti-carbon strategy), biodiversity (the reproduction and cohabitation of living species) and social empathy (the development and promotion of the exchange of social capital – culture, traditions, emotions – among the resident populations).
The Forest City is a vibrant city, with a high proportion of creative relationships. Inhabited by individuals, families, communities and businesses that all experience a completely new sensation of equilibrium between natural and artificial and which use this fertile balance to create innovation in every area of daily life.

森林城市是新生活方式的样板，它直接从垂直森林这一概念延伸而来，是垂直森林的叠加和进化。森林城市是一个垂直城市，它将数百公顷森林里的植物生命汇集到了1～2km²的城市空间之内。森林城市是人类、植物和其他动物的栖息地，它的最低标准是人均2棵树。

森林城市是一个综合复杂的生态系统，它集环境可持续性（低碳战略）、生物多样性（生物繁殖和物种共存）和社会价值（开发和推广社会资本，即城市居民中的文化、传统和情感）为一体。

森林城市是一个充满活力的城市，有很高比例的创新关系。个人、家庭、社区和企业都将体验到自然和人工之间崭新的平衡方式，并且将这有益的平衡应用于日常生活的各个领域，以此创造新的价值。

2.1

How to export VF01
如何输出垂直森林01号

Adapt to local biodiversity.

The Vertical Forest provides a model of combination between the urban and the natural world that can be adapted to different environmental conditions according to the local climatic zones. The vegetative and animal species populating the Vertical Forest in a cold climate will be different from the ones of a temperate or dry or tropical climate. Each climatic zone provides a set of local criteria to adapt the prototype to the local conditions.

适应当地生物多样性

垂直森林为城市与自然界提供了可以结合的模式，根据当地气候带的不同，这个模式可以适用于不同的环境条件。位于寒冷气候带的垂直森林，其中的植物和动物种类与气候温和、或干燥、或热带气候区是不一样的。每一个气候带都有一套适应当地条件的动植物蓝本。

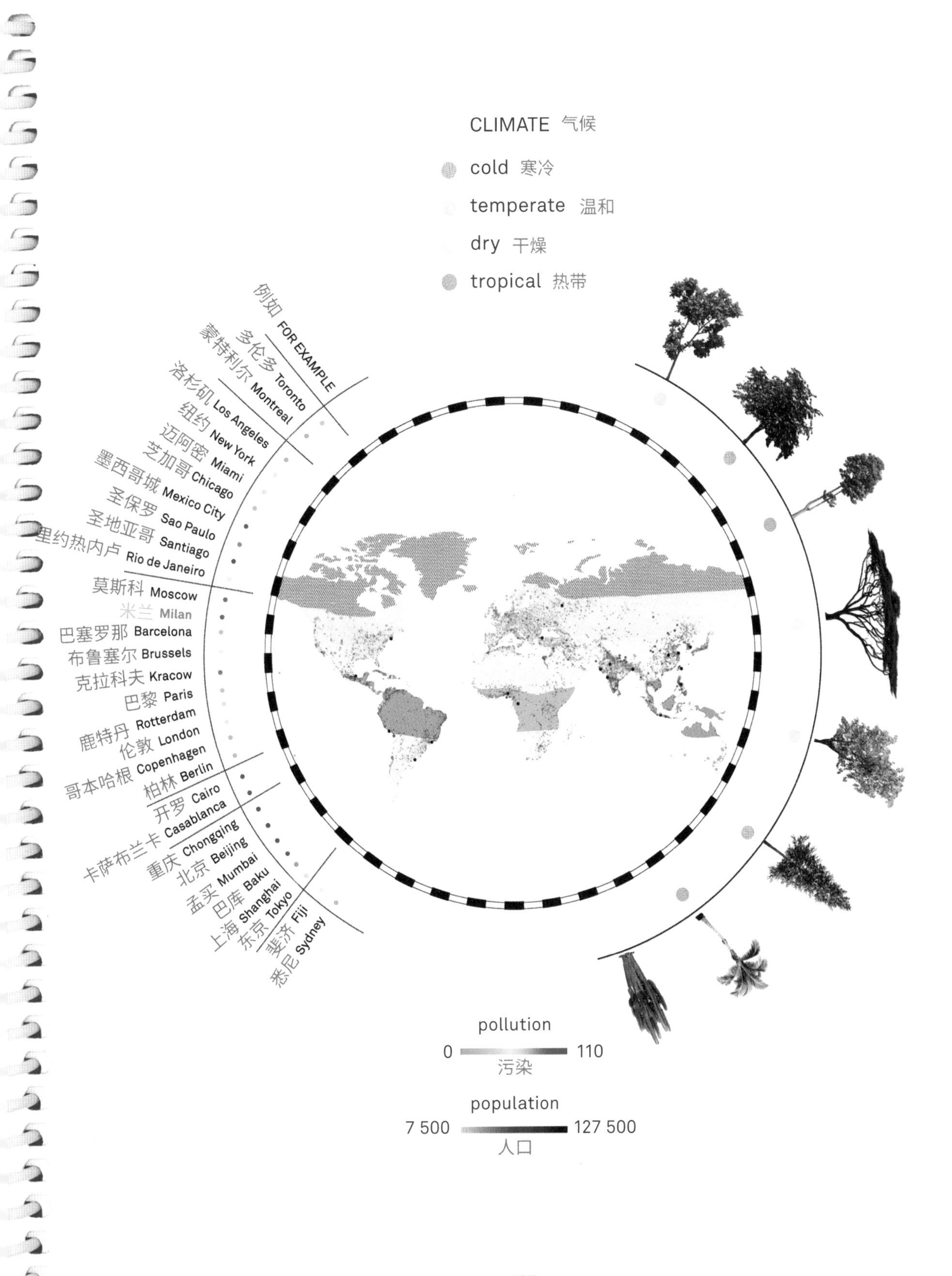
CLIMATE 气候
cold 寒冷
temperate 温和
dry 干燥
tropical 热带
例如 FOR EXAMPLE
多伦多 Toronto
蒙特利尔 Montreal
洛杉矶 Los Angeles
纽约 New York
迈阿密 Miami
芝加哥 Chicago
墨西哥城 Mexico City
圣保罗 Sao Paulo
圣地亚哥 Santiago
里约热内卢 Rio de Janeiro
莫斯科 Moscow
米兰 Milan
巴塞罗那 Barcelona
布鲁塞尔 Brussels
克拉科夫 Kracow
巴黎 Paris
鹿特丹 Rotterdam
伦敦 London
哥本哈根 Copenhagen
柏林 Berlin
开罗 Cairo
卡萨布兰卡 Casablanca
重庆 Chongqing
北京 Beijing
孟买 Mumbai
巴库 Baku
上海 Shanghai
东京 Tokyo
斐济 Fiji
悉尼 Sydney
pollution
0
110
污染
population
7 500
127 500
人口

cold climate　寒冷气候

temperate climate　温和气候

tropical climate　热带气候

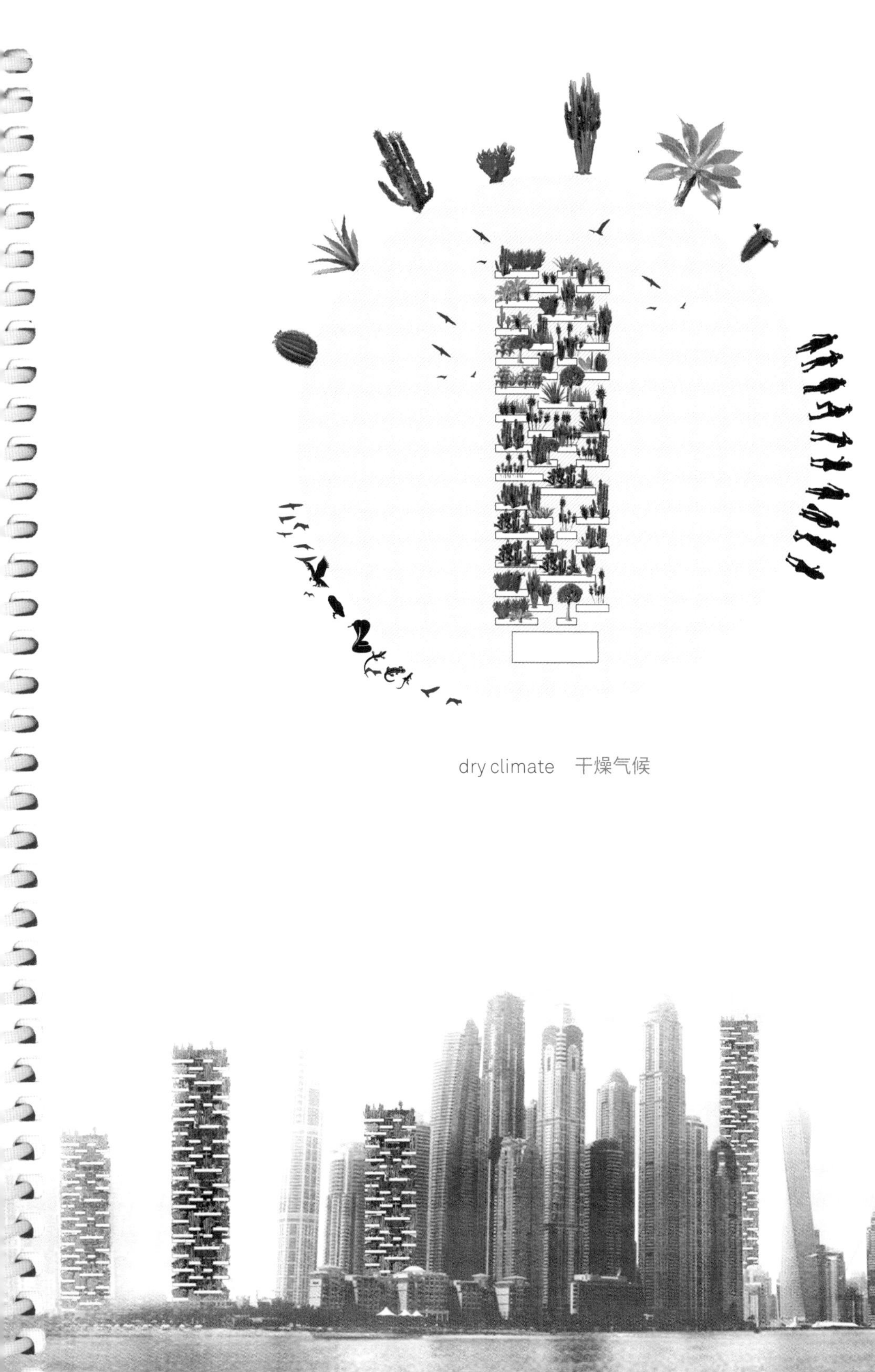

dry climate　干燥气候

2.2

Graft
移植

Greening existing cities.

The Vertical Forest offers a conceptual and practical solution to the increasing environmental crisis by providing the means to intervene in existing cities through the insertion of injections of biodiversity.

Besides the construction of new units of Vertical Forest, biodiversity can also be superimposed on existing structures through designing architectural solutions to host the vegetative elements.

现存城市的绿化

通过在老城中植入生物多样性，垂直森林为日益恶化的环境危机提供了一个兼具概念性和实用性的解决方案。

除了建设新的垂直森林，还可以通过建筑设计方案来聚集各种植物，将生物多样性叠加到已有结构之中。

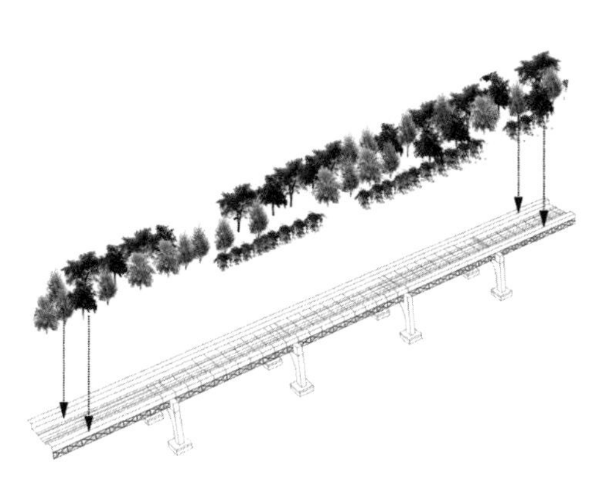

bridge
桥梁

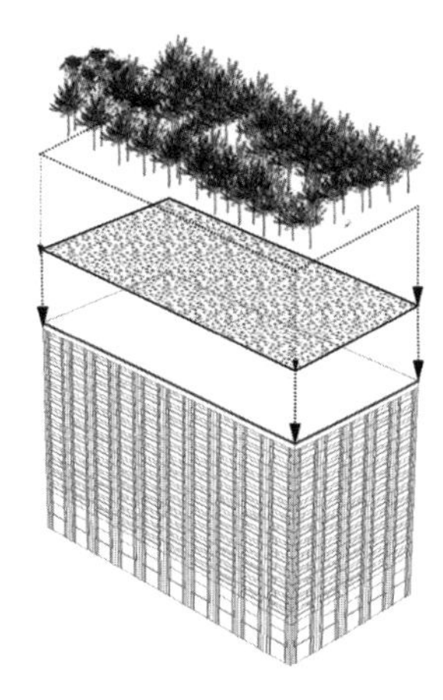

roof
屋顶

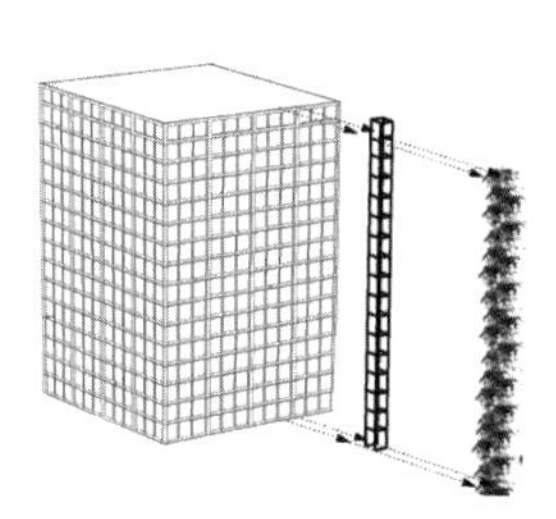

external elevators and stairs
外部电梯和楼梯

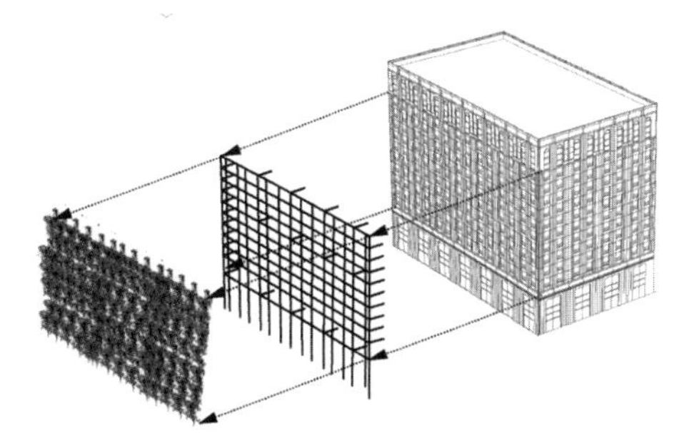

façade
外立面

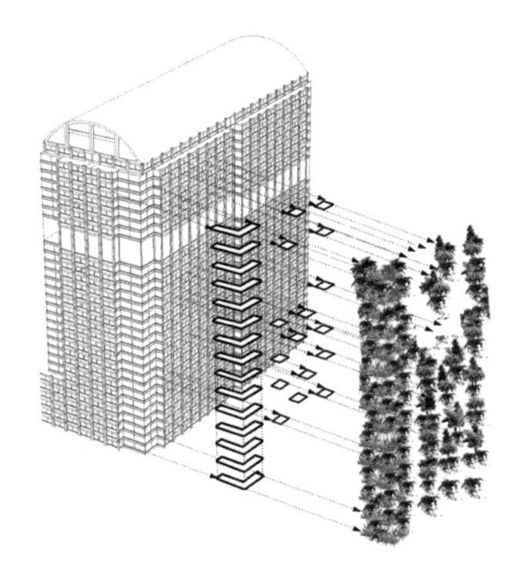

balconies
阳台

Vertical Forest 02
垂直森林02号

In Chavannes-près-Renens the second prototype Vertical Forest is now being realised.
A 36 floors 117 meters high tower that will host 24,000 plants, including 100 trees belonging to the family of cedars. The plant pots are integrated into a system of galleries that redefines the design of the façades.

在勒南附近的沙瓦纳（瑞士），第二个垂直森林的样板已经建成。

一个36层、高117m的塔楼容纳了24 000株植物，其中有100株雪松。所有植物的种植盆被集成为整体系统，并重新赋予建筑立面新的定义。

Credits **项目团队**

Developers and General Contractor **开发和总承包**

Entreprise Générale Bernard Nicod SA – Losanne, CH

Orllati Real Estate SA – Biolay-Orjulaz, CH

Promotion and Management of the procedure **工程监理**

Commune de Chavannes-près-Renens, CH

Architectural Design **建筑设计**

Stefano Boeri Architetti – Milano **博埃里建筑设计事务所**

Partners Stefano Boeri, Michele Brunello

Project leaders Arch. Marco Giorgio, Arch. Marco Bernardini

Team Arch. Moataz Faissal Farid, Arch. Julia Gocalek, Dott. Cécile Cannesson

Agronomist and Landscape Consultant **农学专家和景观顾问**

Studio Laura Gatti – Milano

Engineering Design **结构设计**

BuroHappold Engineering | Building Environments – London

Weinmann-Energies SA – London

2.3

Visions of a Forest City

"The earth", "the gardens" and "the forest"

Within the Forest City the urban, natural and rural worlds (human care and attention to nature) co-exist thanks to the creation of a unique and integrated architectural environment at every level of the settlement.

1. "The earth" is a broad and permeable ground-floor area for productive, commercial and cultural activities. It is a fluid and totally public space, open to free transportation systems and the flow of people, goods and carriers from everywhere.

2. "The gardens" are a horizontal pedestrian platform ranging from 5 to 9 metres above the common space where the inhabitants of the forest and their visitors can meet each other, work and enjoy sports and cultural activities. "The gardens" are a semi-public space that will house vegetable gardens, sports facilities and a carrier system for suspended transport.

3. "The forest" is an authentic forest of skyscrapers ranging from 40 to 200 metres in height, a private landscape where it will be possible to experience a new and unusual closeness between humans, trees and other animal species. "The forest" is a vertical kaleidoscope of internal environments – for human beings who want to live with all the amenities and creature comforts within a genuine forest of trees, plants, bushes, etc.

森林城市的远景

“地面”，“花园”和“森林”

在森林城市中，城市、自然和乡村（人文关怀和关注自然）和谐共存，这得益于一个独特的、综合多层面的建筑环境的营造。

1.“地面”是一个广泛透水的地面区域，以用于生产、商业和文化活动。这是一个流动的、完全公共的空间，向交通系统完全开放，车流、人流、物流无处不达。

2.“花园”是一个水平的人行平台，在公共空间上方5～9m。在那里，森林之城的住户和访客可以见面、互动，享受运动以及文化活动。“花园”是一个半公共空间，它拥有蔬菜园、运动设施以及一个空中的交通系统。

3.“森林”是一组高为40～200m的真实的森林大楼，它是一个私人景观，能够体验人类、植物和其他动物物种之间新的不寻常的亲密接触。

“森林”是室内环境的垂直万花筒，为希望在植物森林环境中享受到所有设施和舒适生活的人们提供去处。

GIUDICE

THE FLYING GARDENERS

飞翔的园丁

Once a year they fly around the Vertical Forest.
They hang by rope from the edge of the roof
and descend by jumping between balconies.
Agronomists and climbers, only they have the consciousness
of the richness of the lives that the Forest hosts
in the Milan sky.

每一年，他们都飞跃在垂直森林周围。

他们从屋顶的边缘放下绳索，下降，跳跃在阳台之间。

他们是农学家和攀登者，只有他们最了解米兰天空下的这片森林中的生命。

The Flying Gardeners is a short-film realised by The Blinkfish during summer 2015, on an idea by Stefano Boeri, in occasion of Chicago Architecture Biennial 2015 and Shanghai Urban Space Art Season. Supported by Comune di Milano, Italian Consulate in USA, ICE, Sisters Cities International and sponsored by Azimut Sgr Spa, Solidea Srl, Chirico Design, Albero della Vita/Orgoglio Brescia.

《飞翔的园丁》是由Blinkfish公司于2015年夏天制作的短剧，基于斯坦法诺·博埃里的构想，在2015芝加哥建筑双年展以及上海城市空间艺术季放映。由米兰市政府、意大利驻美国大使馆、意大利贸易促进会、国际姐妹城市组织支持，由阿资木特总部、 索利德拉公司、奇立可设计所、布雷西亚苗圃赞助。

Stefano Boeri 斯坦法诺·博埃里

A Vertical Forest
一座垂直的森林

edited by / 编辑
圭多·穆桑特 (Guido Musante), 阿族拉·穆重尼格罗(Azzurra Muzzonigro)
with contributions by / 协助
米歇尔·布鲁奈洛 (Michele Brunello), 劳拉·加蒂 (Laura Gatti), 尤丽亚·高卡特赖克 (Julia Gocatek) 和 胥一波
Book design / 装帧设计 Pietro Corraini

texts by / 文字
斯坦法诺·博埃里 (Stefano Boeri), 圭多·穆桑特 (Guido Musante), 阿族拉·穆重尼格罗 (Azzurra Muzzonigro)
translations by / 英文翻译 乔恩·考克斯 (Jou Cox)；
中文翻译 胥一波
photographs by / 图像
保罗·罗塞利 (Paolo Rosselli)，劳拉·茨昂琪 (Laura Cionci) 和 眨眼鱼影像公司 (The Blinkfish)
illustrations by / 插画 左西亚·泽鲁佳富苏卡(Zosia Dzierżawska)
illustrations and diagrams of the Manifesto /
宣言插图：左西亚·泽鲁佳富苏卡(Zosia Dzierżawska), 尤丽亚·高卡特赖克 (Julia Gocalek)，以及博埃里建筑事务所

p. 50 © Emilio Ambasz, Fukuoka Prefectual International Hall for Dai-Chi Mutual Life Insurance Co., Fukuoka, Japan 1990
pp. 53, 88, 90 © Iwan Baan
p. 54 © SITE, *High rise of Homes* - drawing by James Wines - 1981 coll: The Museum of Modern Art, New York
p. 55 Joseph Beuys, *7000 Eichen*: © Joseph Beuys, by SIAE 2015
p. 64 © Ricky Burdett, Urban Age/LSE Cities
p. 64 © Urban Age/LSE Cities
p. 78 © Arup Italia
p. 100 © Gilles Clément, Manifesto del Terzo paesaggio, Quodlibet, Macerata 2005
p. 102 Le Corbusier, Immeubles Villas: *© FLC, by SIAE 2015*

The short film The Flying Gardeners is by The Blinkfish /
《飞翔的园丁》是由眨眼鱼影像公司创作
From an idea of Stefano Boeri / 根据博埃里先生的构想完成
Directors / 导演
贾科莫·博埃里 (Giacomo Boeri)，马特欧·哥立马尔第 (Matteo Grimaldi)
Executive producer / 执行制片人
保罗·索拉维亚 (Paolo Soravia)
Photography Director / 摄影总监
贾科莫·弗利特里 (Giacomo Frittelli)
Gardeners / 参演园丁
吉尔伯特·安托内利 (Gilberto Antonelli)，马西莫·索马尼 (Massimo Sormani)，乔尔瓦尼乌果 (Giovannillgo)

The Vertical Forest is a project by Boeri Studio (Stefano Boeri, Gianandrea Barreca, Giovanni La Varra)
垂直森林由博埃里建筑事务所设计（博埃里，巴莱卡和拉瓦拉）
Supervision of works: Davor Popovic
施工监理：达沃·波波维奇
Landscape project: Laura Gatti and Emanuela Borio
景观工程：劳拉·加蒂，埃曼奴埃拉·博里奥
Supply and maintenance of green: Peverelli s.r.l.
绿化供应与养护：佩维莱立公司
Structures projects: Arup Italia s.r.l.
结构工程：奥雅纳，意大利
Systems design: Deerns Italia S.p.A.
系统设计：Deerns Italia S.p.A.
Asset Manager: COIMA SGR
资产管理：柯意马
Interior design:COIMA Image in collaboration with
室内设计：与柯意马图像公司联合参与
Special thanks to /Kelly Russel, Director of Communication Coima SGR and her team
特别鸣谢：Kelly Russel，柯意马传播总监以及她的团队

The publisher and Stefano Boeri Architetti will be at complete disposal to whom might be related to the unidentified sources printed in this book.
出版社以及博埃里建筑设计事务所将享受该书相关出版权利。

Maurizio Corraini s.r.l.
info@corraini.com
www.corraini.com

同济大学出版社
中国上海四平路1239号
200092
Tel. 8621 6598 5622
www.tongjipress.com.cn